SpringerBriefs in Physics

SpringerBriefs in Physics are a series of slim high-quality publications encompassing the entire spectrum of physics. Manuscripts for SpringerBriefs in Physics will be evaluated by Springer and by members of the Editorial Board. Proposals and other communication should be sent to your Publishing Editors at Springer.

Featuring compact volumes of 50 to 125 pages (approximately 20,000–45,000 words), Briefs are shorter than a conventional book but longer than a journal article. Thus, Briefs serve as timely, concise tools for students, researchers, and professionals.

Typical texts for publication might include:

- A snapshot review of the current state of a hot or emerging field
- A concise introduction to core concepts that students must understand in order to make independent contributions
- An extended research report giving more details and discussion than is possible in a conventional journal article
- A manual describing underlying principles and best practices for an experimental technique
- An essay exploring new ideas within physics, related philosophical issues, or broader topics such as science and society

Briefs allow authors to present their ideas and readers to absorb them with minimal time investment.

Briefs will be published as part of Springer's eBook collection, with millions of users worldwide. In addition, they will be available, just like other books, for individual print and electronic purchase.

Briefs are characterized by fast, global electronic dissemination, straightforward publishing agreements, easy-to-use manuscript preparation and formatting guidelines, and expedited production schedules. We aim for publication 8–12 weeks after acceptance.

More information about this series at http://www.springer.com/series/8902

Jorge Ovalle · Roberto Casadio

Beyond Einstein Gravity

The Minimal Geometric Deformation
Approach in the Brane-World

 Springer

Jorge Ovalle ⓘD
Institute of Physics
Silesian University in Opava
Opava, Czech Republic

Roberto Casadio ⓘD
Dipartimento di Fisica e Astronomia
University of Bologna
Bologna, Italy

ISSN 2191-5423   ISSN 2191-5431   (electronic)
SpringerBriefs in Physics
ISBN 978-3-030-39492-9   ISBN 978-3-030-39493-6   (eBook)
https://doi.org/10.1007/978-3-030-39493-6

This Springer imprint is published by the registered company Springer Nature Switzerland AG
The registered company address is: Gewerbestrasse 11, 6330 Cham, Switzerland

Preface

General Relativity represents one of the greatest intellectual achievements in the history of science. The experimental confirmation of many of its predictions, such as gravitational waves and black holes, place it as one of the pillars of modern physics together with the Standard Model of elementary particles. Despite this success, some fundamental problems still remain without a satisfactory answer in General Relativity, such as the expected formation of space-time singularities in the gravitational collapse and at the beginning of time, as well as the mysterious phenomenology of dark energy and dark matter. These questions have motivated the search for new gravitational theories, or at least reasonable modifications of General Relativity, which could explain our observations and solve the conceptual issues. This book plays in this context, with two main goals: the first one is to show in a clear and simple way, how the possible existence of a (warped) extra spatial dimension would affect the predictions of four-dimensional General Relativity, a model known as the Brane World. While it is true that no experimental evidence has been found in favour of the existence of other dimensions, the model represents an excellent idea which easily explains the hierarchy of fundamental forces, a highly non-trivial fact of nature, and could certainly serve as inspiration in the search for a more realistic theory. The second goal is to explain, step by step, a new technique called Minimal Geometric Deformation, which was introduced in the context of the Brane World, precisely for the purpose of solving the correspondingly modified Einstein field equations. This method has also given rise to the so-called Gravitational Decoupling in General Relativity, which is an extension of the Minimal Geometric Deformation to other scenarios beyond the Brane World. Throughout the book, we place particular emphasis on the study of self-gravitating compact objects, for which all of these methods have originally been introduced and developed. Finally, we would like to highlight that this book can be used as a complement to an advanced course in General Relativity. In particular, it can be used to show how to generate new exact solutions of the Einstein field equations in the interior of self-gravitating sources starting from known ones.

Opava, Czech Republic   Jorge Ovalle
Bologna, Italy   Roberto Casadio
November 2019

Contents

1 The Minimal Geometric Deformation 1
  1.1 The Simplest Extension 2
  1.2 Extra-Dimensional Gravity: The Brane-World 5
     1.2.1 A Brane in the Bulk 6
     1.2.2 Effective Brane-World Equations 7
  1.3 Brane World Non-locality and the General Relativity Limit 11
  1.4 MGD Constraint on the Brane.......................... 13
  1.5 Summary: MGD Procedure in the Brane World 16
  1.6 Matching Conditions for Stellar Distributions 17
  1.7 About Exact Solutions 21
  References 24

2 Stellar Distributions 27
  2.1 The Heintzmann Solution 28
     2.1.1 Matching Conditions 30
  2.2 The Interior Schwarzschild Solution 34
     2.2.1 The Interior Weyl Fluid 36
     2.2.2 Matching Conditions 38
  2.3 The Durgapal-Fuloria Solution 40
     2.3.1 Matching Conditions 42
  2.4 Tolman IV Solution 44
     2.4.1 General Relativity Limit......................... 47
     2.4.2 Matching Conditions 47
  2.5 Brane World Stars with Solid Crust...................... 49
     2.5.1 Stellar Interior 50
     2.5.2 Energy Conditions 53
     2.5.3 The Solid Crust 56
  2.6 Classical Tests of General Relativity 57
     2.6.1 Perihelion Precession 59
     2.6.2 Light Deflection 60

		2.6.3	Radar Echo Delay	61
		2.6.4	Further Remarks	62
	References			63

3 Microscopic Black Holes 65
   3.1 Regular Brane-World Stars and Black Holes 66
   3.2 Exterior Geometry 67
   3.3 Interior Geometry 69
   3.4 Black Hole Limit and Minimum Mass 70
      3.4.1 More Linearly Charged Brane-World Black Holes 72
      3.4.2 Other Interior Geometries 75
   References 76

4 A Generalization of the Minimal Geometric Deformation 77
   4.1 The Exterior Geometry 78
   4.2 Extended Geometric Deformation 79
   4.3 Modified Exterior Solution 81
   4.4 Five-Dimensional Solutions 83
   4.5 New Exact Solutions 86
      4.5.1 Outer Solutions 88
      4.5.2 The Interior 91
   References 92

5 Gravitational Decoupling 95
   5.1 MGD Decoupling for Two Sources 95
   5.2 Gravitational Decoupling in the Brane-World 99
      5.2.1 Black Holes 101
      5.2.2 Interior Solutions 104
      5.2.3 Beyond Perfect Fluid Solutions by MGD Decoupling
           in the Brane-World 105
   References 110

Acronyms

ADM	Arnowitt-Deser-Misner
BW	Brane World
DEC	Dominant Energy Condition
DMPR	Dadhich-Maartens-Papadopoulos-Rezania
FLRW	Friedmann-Lemaître-Robertson-Walker
GD	Geometric Deformation
GR	General Relativity
LHC	Large Hadron Collider
MGD	Minimal Geometric Deformation
NEC	Null Energy Condition
RS	Randall-Sundrum
SEC	Strong Energy Condition
WEC	Weak Energy Condition
WMAP	Wilkinson Microwave Anisotropy Probe

Chapter 1
The Minimal Geometric Deformation

General Relativity (GR) in its century of existence represents, without a doubt, one of the most important achievements of human knowledge. The predictions made by this theory, like the perihelion shift of Mercury, light deflection and gravitational lensing, the gravitational redshift and time delay, black holes and gravitational waves—just to mention some of the most remarkable ones, have given it the honours which deserves as one of the fundamental theories of Physics (for a recent review on experimental tests of GR see [1] and references therein). On the other hand, the great technological advances during the past decades have provided us with increasingly powerful instruments which allow for accurate measurements of the evolution of compact systems, which represent the best candidate laboratories for the study of gravity in the strong regime (see for instance [2]). This progress not only serves to test the theory like never before, but also appears to leave GR as the most reliable gravitational theory to be used in the analysis of phenomena occurring in the strong field regime. Likewise, the ability of observing increasingly distant objects deep in the universe, thus with a larger and larger cosmological redshift, leads to the conclusion that we can obtain an excellent modelling of these phenomena within GR [3, 4]. Additionally, due to the results obtained by the PLANCK collaboration [5], which improved greatly on the previous findings by WMAP [6], the cosmological models based on GR are enjoying a well-deserved and well-earned recognition. Furthermore, the recent direct observation of black holes shadows by the EHT collaboration [7, 8], as well as the detection of gravitational waves by the LIGO and Virgo collaborations [9–11], allows us to ensure that GR is one of the most successful scientific theories in the history of science. Why would we want to find new gravitational theories beyond GR then?

Despite the above success, there are fundamental questions about the gravitational interaction which GR cannot answer satisfactorily. These can be broadly grouped in the following fundamental issues, which could likely be closely related:

© The Author(s), under exclusive license to Springer Nature Switzerland AG 2020
J. Ovalle and R. Casadio, *Beyond Einstein Gravity*, SpringerBriefs in Physics,
https://doi.org/10.1007/978-3-030-39493-6_1

(1) The inability of GR to explain satisfactorily the phenomenology of dark matter [12, 13] and dark energy, without introducing some kind of unknown matter-energy to reconcile its predictions with the observed galactic rotation curves and accelerated expansion of the universe; (2) The general GR prediction of the existence of space-time singularities, most notably in the gravitational collapse of regular matter distributions, which are both mathematically and physically hard to accept; (3) The impossibility, so far, to reconcile GR with the Standard Model of particle physics, or equivalently, the difficulties met when one tries and quantize GR by the same schemes used successfully for the other fundamental interactions. All of these have strongly motivated searches for a gravitational theory beyond GR. If the new theory is a consistent quantum theory, this could lead to a modification of GR at low energy which accounts for the dark matter and dark energy. If the new theory is not a consistent quantum theory for gravity, this should also contain GR in a suitable limit, and somehow show greater tolerance to its quantum description.

Indeed, there are many alternative theories, as for instance, high curvature gravity theories, Galileon theories, $f(R)$ theories, scalar-tensor theories, massive gravity, new massive gravity, topologically massive gravity, Chern-Simons theories, higher spin gravity theories, Horava-Lifshitz gravity, extra-dimensional theories, etc. All of these proposals entail a modification of the original Einstein equations

$$G_{\mu\nu} \equiv R_{\mu\nu} - \frac{1}{2} R \, g_{\mu\nu} = k^2 \, T_{\mu\nu} \,, \tag{1.1}$$

where Greek indices run from 0 to 3, $R_{\mu\nu}$ and R are, respectively, the Ricci tensor and its trace determined by the space-time metric $g_{\mu\nu}$, $T_{\mu\nu}$ is the energy-momentum tensor of the matter which sources gravity, and $k^2 = 8\,\pi\,G_{\rm N}/c^2$ with $G_{\rm N}$ the Newton constant (we shall also mostly set the speed of light $c = 1$ and use the metric signature $+ - --$, unless otherwise specified).

1.1 The Simplest Extension

Any extension of GR will eventually give rise to new terms in the effective four-dimensional Einstein equations (1.1). These "corrections" can usually be cast in the form of an effective energy-momentum tensor and appear in such a way that they should vanish or be negligible in an appropriate limit, as for instance, at solar system scale, where GR has been successfully tested. This limit represents not only a critical point when a consistent extension to GR is studied, but also a non-trivial problem that must be treated carefully.

The simplest way to produce an extension to GR is by considering a modification of the Hilbert-Einstein action as

$$S = \int \left(\frac{R}{2\,k^2} + \mathcal{L}_M \right) \sqrt{-g} \, d^4x + \alpha \, (correction) \,, \tag{1.2}$$

where α is a free parameter associated with the new gravitational sector not described by GR. The explicit form corresponding to the generic correction shown in Eq. (1.2) should be, of course, a well justified and physically motivated expression. At this stage, the GR limit obtained by setting $\alpha = 0$, is just a trivial issue, and everything looks consistent so far. Indeed, we may go further and calculate the equations of motion corresponding to this new theory. Upon demanding as usual that the variation $\delta S = 0$, one obtains

$$R_{\mu\nu} - \frac{1}{2} R\, g_{\mu\nu} = k^2\, T_{\mu\nu} + \alpha\, (new\ term)_{\mu\nu}\ , \tag{1.3}$$

where the new term in Eq. (1.3) may be consolidated as part of an effective energy-momentum tensor. The explicit form of this tensor may contain some new degrees of freedom, as for instance scalar, vector and tensor fields, all of them coming from the new gravitational sector not described by Einstein's original theory. The GR limit is still a trivial issue, since $\alpha = 0$ leads to the standard Einstein equations (1.1).

The above arguments seem to tell us that the consistency with the GR predictions—more formally, the GR limit—is a trivial problem. However, when the system of Eq. (1.3) is solved, the result may show a completely different story. In fact, it is very common that a general solution of Eq. (1.3) does not reproduce any GR space-time metric by simply setting $\alpha = 0$. The origin of this problem is in the non-linearity of the system (1.3), and it should not come as a surprise. To clarify this point, let us consider a static and spherically symmetric perfect fluid. The most general spherically symmetric space-time metric can be written as

$$\begin{aligned} ds^2 &= g_{\mu\nu}\, dx^\mu\, dx^\nu \\ &= e^{\nu(r)}\, dt^2 - e^{\lambda(r)}\, dr^2 - r^2 \left(d\theta^2 + \sin^2\theta\, d\phi^2\right)\ , \end{aligned} \tag{1.4}$$

where ν and λ are functions of the areal radius r only. The Einstein equations (1.1) then yield the metric component

$$e^{-\lambda} = 1 - \frac{2\, m(r)}{r}\ , \tag{1.5}$$

where

$$m = \frac{k^2}{2} \int_0^r \rho(x)\, x^2\, dx \tag{1.6}$$

is the (Misner-Sharp) mass function of the self-gravitating system. Now let us consider the same perfect fluid under the "new" gravitational theory defined in Eq. (1.2). When the Eq. (1.3) are solved we obtain an expression which may generically be written as

Fig. 1.1 MGD: the new
gravitational sector not
described by GR is fully
controlled by the
deformation parameter α.
GR represents a "sub-space"
in this extended version of
gravity and is recovered for
$\alpha = 0$

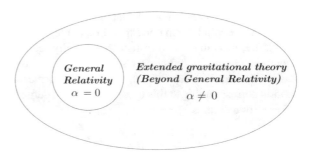

$$e^{-\lambda} = 1 - \frac{2\, m(r)}{r} + (geometric\ deformation) \,, \qquad (1.7)$$

where the term *geometric deformation* should be understood as denoting the modification undergone by the metric component in Eq. (1.5) due to the generic extension of GR shown in Eq. (1.2). Therefore "deformation" means the difference with respect to a specific GR solution. Now, and this is very important, the deformation shown in the expression (1.7) always induces an *effective anisotropy* on the perfect isotropic fluid. Hence, the self-gravitating system does not behave like a perfect fluid anymore. Indeed, and this is a critical point in our analysis, the anisotropy $\mathcal{P}$ produced by the geometric deformation always appears to be of the form

$$\mathcal{P} = A + \alpha\, B \,. \qquad (1.8)$$

This expression shows that the GR limit cannot be recovered simply by setting $\alpha = 0$, since the α-independent contribution A in the anisotropy (1.8) will survive in this limit. Consequently, the perfect fluid GR solution (1.5) is not contained in this extension and we could say that we have an extension to GR which does not contain GR. This seems a contradiction, but it is in fact mathematically correct and appears more properly as a problem of physical consistency. The source of this problem is the *geometric deformation* shown in Eq. (1.7), which always appears as

$$geometric\ deformation = X + \alpha\, Y \,. \qquad (1.9)$$

That is, there is a "sector" of the geometric deformation, generically called X, which is α-independent. Obviously this is physically problematic, since we expect to be able to control the deformation undergone by GR by varying the parameter α. It is important to remark once more that the source of this problem is solely given by the high non-linearity of the effective Einstein equations (1.3), and has nothing to do with any specific extensions of GR.

Fig. 1.2 MGD: when a GR
solution is kept as a solution
in the new gravitational
theory, the α-independent
new terms in the extended
solution vanish, and the GR
limit can be recovered

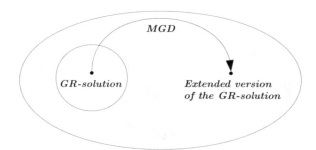

A method devised to solve precisely the non-trivial issue of consistency described above is the so-called *Minimal Geometric Deformation approach* (MGD) [14, 15]. The key idea is to keep under control the anisotropy induced by the extension to the Einstein theory. In particular, if we demand that the α-independent sector X in the geometric deformation (1.9) vanishes identically, the α-independent contribution to the anisotropy A in Eq. (1.8) will also vanish as a consequence. This procedure ensures that the extended theory will recover GR in the limit $\alpha = 0$. In this approach the generic expression Y in Eq. (1.9) represents the *minimal geometric deformation* undergone by the radial metric component, being the generic expression B in Eq. (1.8) the *minimal anysotropy* added to GR by the correction terms in the action (1.2) (see Figs. 1.1, 1.2).

Now, the question is how to implement the approach described above in practice. Or, more clearly, how to impose $X = 0$ in Eq. (1.9) in order to obtain a consistent extension to GR? This is accomplished when a GR solution is forced to remain a solution in the extended theory. Roughly speaking, we need to introduce the GR solution into the new theory, as far as possible. This provides the practical foundation for the MGD approach [16–24]. We want to emphasise that the GR solution used to make $X = 0$ in Eq. (1.9) will eventually be modified by using, for instance, the matching conditions at the surface of compact astrophysical object. This will in general allow us to express physical variables as functions of the deformation parameter of the theory, here generically named α. The first application of the MGD was to so called Brane-World (BW) models, and the parameter α was therefore given by the (inverse of the) brane tension. For this reason, most of this book will consider BW models of stars and other compact objects, such as black holes, although the MGD can be applied to any modification of GR of the form described above.

1.2 Extra-Dimensional Gravity: The Brane-World

Superstring/M-theory has been long considered one of the most promising candidates of quantum gravity. It describes gravity as a higher-dimensional interaction which becomes effectively four-dimensional at low enough energies. Extra-dimensional

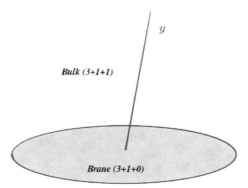

Fig. 1.3 BW: space-time is five-dimensional with a spatial extra-dimension y. The Standard Model is trapped in the four-dimensional brane, while gravity can propagate in the complete five-dimensional space (the bulk). The hierarchy problem is solved in terms of a fundamental five-dimensional coupling near the electro-weak scale

theories are among the theories that lead to modifications to GR. One of these extra-dimensional theories is the five-dimensional BW proposed by Randall and Sundrum (RS) [25, 26], which has been largely studied and which can explain rather straight-forwardly the hierarchy problem in the Standard Model of particle physics (see also the ADD model [27, 28]). In this book we shall only consider the BW modifications to GR. For this purpose, some fundamental aspects will be mentioned, but no detailed account of the BW theory will be given. The interested reader can find excellent reviews, for example, in the works of Maartens and Koyama [29, 30].

1.2.1 A Brane in the Bulk

In BW models, the gauge and fermion fields of the Standard Model are confined to our observable universe (the brane), while gravity propagates in all (four) spatial dimensions (the bulk). Consequently, the fundamental gravitational scale κ_5 is reduced to the weak scale by considering extra-dimensional effects, and the weakness of gravity relative to the other forces is an apparent effect due to the warping along the extra spatial direction. Because of this, its study and impact on GR is of great conceptual importance, although no experimental evidence to support it has been found [31, 32]. However, we should mention that a good agreement was found between the BW theoretical predictions and the dark matter observations in Ref. [33].

It is mostly convenient to describe the space-time around the brane by employing Gaussian normal coordinates $x^A = (x^\mu, y)$,[1] where y is the extra-dimensional coordinate such that our brane is located at $y = 0$ (see Fig. 1.3). The five-dimensional

---

[1] Capital letters run from 0 to 4, whereas Greek letters run from 0 to 3 as usual.

metric can then be expanded near the brane as

$$g^5_{AB} \simeq g^5_{AB}\big|_{y=0} + 2\,K_{AB}\big|_{y=0}\ y + \pounds_{\hat{n}} K_{AB}\big|_{y=0}\ y^2\ , \qquad (1.10)$$

where K_{AB} is the extrinsic curvature of the brane, and $\pounds_{\hat{n}}$ the Lie derivative along the unitary four-vector $\hat{n}$ orthogonal to the brane. Junction conditions at the brane lead to [34]

$$K_{\mu\nu} \sim T_{\mu\nu} - \frac{1}{3}\ (T - \sigma)\ g_{\mu\nu}\ , \qquad (1.11)$$

where σ is the brane tension (or vacuum energy, from the four-dimensional perspective) and $T_{\mu\nu}$ is the stress tensor of the Standard Model matter confined on the brane. Moreover, one also has

$$\pounds_{\hat{n}} K_{\mu\nu} \sim \mathcal{E}_{\mu\nu} + F_{\mu\nu}\ , \qquad (1.12)$$

where $\mathcal{E}_{\mu\nu}$ is the projection of the Weyl tensor on the brane and $F_{\mu\nu}$ a tensor which depends on $T_{\mu\nu}$ and σ.

We are primarily interested in describing stars and other compact objects on the brane. It is therefore important to recall that the junction conditions in GR [35] allow for (step-like) discontinuities in the stress tensor $T_{\mu\nu}$, keeping the first and second fundamental forms continuous. For thin (Dirac δ-like) surfaces, a step-like discontinuity of the extrinsic curvature $K_{\mu\nu}$ orthogonal to the surface is also allowed if the metric remains continuous [35]. Since the brane in the RS model is itself a thin surface, it should generate an orthogonal discontinuity of the extrinsic curvature K_{AB} in five dimensions. Consequently, discontinuities in the localised matter stress tensor $T_{\mu\nu}$ would induce discontinuities in the extrinsic curvature (1.11) tangential to the brane, which should appear in the five-dimensional metric (1.10). Such discontinuities of the metric g_{AB} are not allowed by the regularity of five-dimensional geodesics. Moreover, because of the second order term in Eq. (1.10), and considering Eq. (1.12), the projected Weyl tensor should not be discontinuous on the brane either. Of course, in a microscopic description of the BW, matter should be smooth along the extra dimension, yet localised on the brane (say, within a width of order $\sigma^{-1/2}$, see e.g. Ref. [36]). In any such description, the continuity of five-dimensional geodesics must then hold and, in order to build a physical model of a star, one has to smooth both the matter stress tensor and the projected Weyl tensor across the surface of the star along the brane.

1.2.2 Effective Brane-World Equations

The general five-dimensional Einstein equations with a bulk cosmological constant Λ_5 read

$$R_{AB} - \frac{1}{2}\ R\,g_{AB} + \Lambda_5\,g_{AB} = \kappa_5\,T_{AB}\ , \qquad (1.13)$$

where κ_5 denotes the five-dimensional gravitational coupling. These equations can be projected onto the four-dimensional brane at $y = 0$ to give the modified four-dimensional Einstein equations [34]

$$R_{\mu\nu} - \frac{1}{2} R\, g_{\mu\nu} + \Lambda\, g_{\mu\nu} = k^2\, T_{\mu\nu}^{\text{eff}} \, , \tag{1.14}$$

where

$$8\,\pi\, G_{\text{N}} = k^2 = \frac{1}{6}\,\sigma\,\kappa_5^4 \, , \tag{1.15}$$

and Λ is the four-dimensional cosmological constant. The four-dimensional energy-momentum tensor $T_{\mu\nu}$ is complemented by new terms, which act as an effective matter source, namely

$$T_{\mu\nu}^{\text{eff}} = T_{\mu\nu} + \frac{6}{\sigma}\, S_{\mu\nu} + \frac{1}{k^2}\, \mathcal{E}_{\mu\nu} + \frac{4}{\sigma}\, \mathcal{F}_{\mu\nu} \, , \tag{1.16}$$

where σ is again the brane tension. The term $S_{\mu\nu}$ represents a high-energy correction quadratic in the energy-momentum tensor ($T = T_\alpha{}^\alpha$), and is given by

$$S_{\mu\nu} = \frac{1}{4}\, T_{\mu\alpha}\, T^\alpha{}_\nu - \frac{T}{12}\, T_{\mu\nu} + \frac{1}{24}\, g_{\mu\nu}\left(T^2 - 3\, T_{\alpha\beta}\, T^{\alpha\beta}\right) \, , \tag{1.17}$$

whereas $\mathcal{E}_{\mu\nu}$ is the (non-local) Kaluza-Klein correction induced by the (in general non-vanishing) five-dimensional Weyl curvature. By associating to this term an effective fluid of four-velocity u^μ, one can write

$$k^2\, \mathcal{E}_{\mu\nu} = \frac{6}{\sigma}\left[\mathcal{U}\left(u_\mu\, u_\nu - \frac{1}{3}\, h_{\mu\nu}\right) + \mathcal{P}_{\mu\nu} + \mathcal{Q}_{(\mu}\, u_{\nu)}\right] \, , \tag{1.18}$$

where $\mathcal{U}$ is the bulk Weyl scalar, $\mathcal{P}_{\mu\nu}$ the effective fluid stress tensor and $\mathcal{Q}_\mu$ the energy flux (the tensor $h_{\mu\nu} = g_{\mu\nu} - u_\mu u_\nu$ projects orthogonally to the fluid lines).

The extra terms $\mathcal{F}_{\mu\nu}$ contains contributions from all (non-standard model) fields possibly living in the bulk, and whose energy-momentum tensor is given by T_{AB} in Eq. (1.13). The five-dimensional cosmological constant Λ_5 is usually fine-tuned to σ in order to generate a small four-dimensional cosmological constant Λ. Indeed, there is a well known relation between the effective brane four-dimensional cosmological constant Λ, the bulk cosmological constant Λ_5, and the brane tension σ, provided by the fine tuning [30]

$$\Lambda = \frac{\kappa_5^2}{2}\left(\Lambda_5 + \frac{1}{6}\,\kappa_5^2\,\sigma^2\right) \, . \tag{1.19}$$

In this book, we shall consider only a cosmological constant in the bulk (which, by fine tunning, will produce $\Lambda = 0$ on the brane), hence

$$\mathcal{F}_{\mu\nu} = 0 \,, \tag{1.20}$$

which implies the conservation equation

$$\nabla_\mu T^{\mu\nu} = 0 \,, \tag{1.21}$$

so that there will be no exchange of energy between the bulk and the brane.

A very particular case is given by the effective four-dimensional equations where $T_{\mu\nu} = 0$, that is

$$R_{\mu\nu} - \frac{1}{2} R\, g_{\mu\nu} = \mathcal{E}_{\mu\nu} \,. \tag{1.22}$$

Since $\mathcal{E}^\mu_{\ \mu} = 0$. The above yield

$$R = 0 \,, \tag{1.23}$$

which is what one usually refers to as the BW vacuum equation.

We shall only consider spherically symmetric static distributions, hence

$$\mathcal{Q}_\mu = 0 \tag{1.24}$$

and

$$\mathcal{P}_{\mu\nu} = \mathcal{P}\left(r_\mu r_\nu + \frac{1}{3} h_{\mu\nu}\right) \,, \tag{1.25}$$

where r_μ is a unit radial vector. Let us then restrict to spherical symmetry and choose as the source term in Eq. (1.14) a perfect fluid,

$$T_{\mu\nu} = (\rho + p)\, u_\mu u_\nu - p\, g_{\mu\nu} \,, \tag{1.26}$$

where $u^\mu = e^{-\nu/2}\, \delta^\mu_0$ is the fluid 4-velocity field in the Schwarzschild-like coordinates of the metric (1.4), with the areal coordinate r ranging from $r = 0$ (the star's centre) to some $r = R_s$ (the star's surface). This metric must then satisfy the effective Einstein field equations (1.14), which, with all the simplifications described above, explicitly read

$$k^2 \left[\rho + \frac{1}{\sigma}\left(\frac{\rho^2}{2} + \frac{6}{k^4}\mathcal{U}\right)\right] = \frac{1}{r^2} - e^{-\lambda}\left(\frac{1}{r^2} - \frac{\lambda'}{r}\right) \tag{1.27}$$

$$k^2 \left[p + \frac{1}{\sigma}\left(\frac{\rho^2}{2} + \rho\, p + \frac{2}{k^4}\mathcal{U}\right) + \frac{4}{k^4}\frac{\mathcal{P}}{\sigma}\right] = -\frac{1}{r^2} + e^{-\lambda}\left(\frac{1}{r^2} + \frac{\nu'}{r}\right) \tag{1.28}$$

$$k^2 \left[p + \frac{1}{\sigma}\left(\frac{\rho^2}{2} + \rho\, p + \frac{2}{k^4}\mathcal{U}\right) - \frac{2}{k^4}\frac{\mathcal{P}}{\sigma}\right]$$
$$= \frac{1}{4}e^{-\lambda}\left(2\nu'' + \nu'^2 - \lambda'\nu' + 2\frac{\nu' - \lambda'}{r}\right) \,, \tag{1.29}$$

where $f' \equiv \partial_r f$. Moreover, the conservation equation (1.21) implies

$$p' = -\frac{\nu'}{2}(\rho + p) \,. \tag{1.30}$$

We then note that the four-dimensional GR equations are formally recovered for $\alpha \equiv 1/\sigma \to 0$, and the conservation equation (1.30) then becomes a linear combination of Eqs. (1.27)–(1.29).

By simple inspection of the field equations (1.27)–(1.29), we can identify the effective density

$$\tilde{\rho} = \rho + \frac{1}{\sigma}\left(\frac{\rho^2}{2} + \frac{6}{k^4}\mathcal{U}\right) \,, \tag{1.31}$$

the effective radial pressure

$$\tilde{p}_r = p + \frac{1}{\sigma}\left(\frac{\rho^2}{2} + \rho\, p + \frac{2}{k^4}\mathcal{U}\right) + \frac{4\,\mathcal{P}}{k^4\,\sigma} \,, \tag{1.32}$$

and the effective tangential pressure

$$\tilde{p}_t = p + \frac{1}{\sigma}\left(\frac{\rho^2}{2} + \rho\, p + \frac{2}{k^4}\mathcal{U}\right) - \frac{2\,\mathcal{P}}{k^4\,\sigma} \,. \tag{1.33}$$

This result clearly illustrate that extra-dimensional effects produce anisotropies in the stellar distribution, given by

$$\Pi \equiv \tilde{p}_r - \tilde{p}_t = \frac{6\,\mathcal{P}}{k^4\,\sigma} \,. \tag{1.34}$$

The isotropic perfect fluid one would mostly use in GR to model the distribution of matter inside a star therefore becomes an anisotropic stellar system on the brane.

The system of Eqs. (1.27)–(1.30) contains six unknown functions, namely: two physical variables, the density ρ and pressure p; two geometric functions, the temporal metric function ν and the radial one λ; and two extra-dimensional fields, the Weyl scalar function $\mathcal{U}$ and the anisotropy $\Pi \sim \mathcal{P}$. Hence it represents an indefinite system of equations on the brane, an open problem for which the solution requires more information about the bulk geometry and a better understanding of how our four-dimensional space-time is embedded in the bulk [37–40]. Since the source of this problem is directly related with the projection $\mathcal{E}_{\mu\nu}$ of the bulk Weyl tensor on the brane, the first logical step to overcome this issue would be to discard the cause of the problem, namely, to impose the constraint $\mathcal{E}_{\mu\nu} = 0$ on the brane. However it was shown in Ref. [41] that this condition is incompatible with the Bianchi identity on the brane, thus a different and less radical restriction must be implemented. In this respect, a useful path that has been successfully used consists in discarding the anisotropic stress associated to $\mathcal{E}_{\mu\nu}$, that is to set $\mathcal{P}_{\mu\nu} = 0$. However this constraint, which is useful to overcome the non-closure problem [42], represents too strong a restriction, since some anisotropic effects onto the brane should be expected as a consequence of the "deformation" undergone by the four-dimensional geometry due to five-dimensional gravity, as was clearly explained in Ref. [43].

1.3 Brane World Non-locality and the General Relativity Limit

We shall see that a useful constraint arises on the brane when the GR limit is required to hold for any BW solution. There is a particular solution to this constraint, a constraint itself on the brane, the physical meaning of which represents a condition of minimal anisotropy projected onto the brane. We shall show that this condition not only ensures a correct low energy limit, but it also represents a condition that is satisfied for *any* known GR solution. The main goal of this section is to show that, on demanding the minimal anisotropic effects onto the brane, as shall be explained in detail, it is possible to construct the BW version of every GR solution. This overcomes the non-closure problem of the BW equations and induces a natural BW generalisation for any GR solution.

The first step is to rewrite the field equations (1.27)–(1.29) as follows: on using the effective density (1.31), the first order differential equation (1.27) can be formally solved by

$$e^{-\lambda} = 1 - \frac{k^2}{r} \int_0^r x^2 \left[\rho + \frac{1}{\sigma} \left(\frac{\rho^2}{2} + \frac{6}{k^4} \mathcal{U} \right) \right] dx .$$

Next, by combining the field equations (1.28) and (1.29), we determine the Weyl fields as

$$\Pi = \frac{6 \mathcal{P}}{k^2 \sigma} = G^1{}_1 - G^2{}_2 ,$$

$$\frac{6 \mathcal{U}}{k^4 \sigma} = -\frac{3}{\sigma} \left(\frac{\rho^2}{2} + \rho\, p \right) + \frac{1}{k^2} \left(2\, G^2{}_2 + G^1{}_1 \right) - 3\, p ,$$

with

$$G^1{}_1 = -\frac{1}{r^2} + e^{-\lambda} \left(\frac{1}{r^2} + \frac{\nu'}{r} \right) ,$$

and

$$G^2{}_2 = \frac{1}{4} e^{-\lambda} \left(2\, \nu'' + \nu'^2 - \lambda'\, \nu' + 2 \frac{\nu' - \lambda'}{r} \right) .$$

In the case of a non-uniform static distribution with local terms (high energy corrections) and non-local terms (bulk Weyl curvature contributions), Eqs. (1.35)–(1.39) form an indefinite system of equations on the brane for the set of three unknown functions $\{p(r), \rho(r), \nu(r)\}$ satisfying the conservation Eq. (1.30). Hence, in order to obtain a solution, we must add additional information. However, as mentioned earlier, it is not clear what kind of restriction should be considered to close the system on the brane. So far, the only thing we know is that the non-closure of the BW equations is directly related with the projection of the bulk Weyl tensor $\mathcal{E}_{\mu\nu}$ on the brane.

To clarify how one can obtain some criterion to help in searching for a solution to the problem described in the previous paragraph, let us start with the apparently simplest way to find a solution to the system of equations on the brane: on substituting Eq. (1.37) into the field Eq. (1.27), we obtain a first order linear differential equation for the geometric function $\lambda = \lambda(r)$, to wit

$$e^{-\lambda}\left(\lambda' - \frac{\nu'' + \nu'^2/2 + 2\nu'/r + 2/r^2}{\nu'/2 + 2/r}\right) = k^2 \frac{\rho - 3\,p - \sigma^{-1}\rho\,(\rho + 3p)}{\nu'/2 + 2/r}$$
$$- \frac{2}{r^2(\nu'/2 + 2/r)}\,, \qquad (1.40)$$

whose formal solution is given by

$$e^{-\lambda} = e^{-I}\left(\int_0^r \frac{e^I}{\nu'/2 + 2/x}\left\{\frac{2}{x^2} - k^2\left[\rho - 3\,p - \frac{\rho}{\sigma}\,(\rho + 3\,p)\right]\right\}dx + c\right)\,, \qquad (1.41)$$

with c an integration constant and

$$I \equiv \int^r \frac{\nu'' + \nu'^2/2 + 2\nu'/x + 2/x^2}{\nu'/2 + 2/x}\,dx\,. \qquad (1.42)$$

Given a solution with p, ρ and ν to (1.30), we would be able to find λ, $\mathcal{P}$ and $\mathcal{U}$ from (1.41), (1.36) and (1.37) respectively. Therefore, from the point of view of a brane observer, finding a solution seems not very complicated, at least from the mathematical point of view. However, it was shown in Ref. [14] that finding a consistent solution by starting from any arbitrary solution $\{p, \rho, \nu\}$ to the conservation Eq. (1.30), in general does not lead to a solution for λ having the expected form (1.7), that is

$$e^{-\lambda} = 1 - \frac{k^2}{r}\int_0^r \rho x^2\,dx + \frac{1}{\sigma}\,(bulk\ effects)\,. \qquad (1.43)$$

In turn, if the solution cannot be written like in Eq. (1.43), the GR solution will not be recovered in the limit $1/\sigma \to 0$. Unfortunately, this precisely happens when we start from an arbitrary solution $\{p, \rho, \nu\}$ to (1.30). The source of this problem has to do with the formal solution for λ given in Eq. (1.41), which mixes GR terms with non-local bulk terms in such a way that makes it impossible to recover GR straightforwardly.

For instance, the following expressions

$$\rho = A + B\,r^{q/2}\,, \quad p = -A - \frac{B}{2}\,r^{q/2}\,, \quad e^{\nu/2} = C\,r^{q/2} \tag{1.44}$$

solve the conservation equation (1.30) for $q > 0$, and with A, B and C constants. However, the radial metric component obtained from Eq. (1.41) cannot be written in the form (1.43), and therefore the expected low energy limit (GR) is not recovered for $1/\sigma \to 0$.

A different way to explain why the formal solution (1.41) leads to the so-called "GR limit problem" is detailed next. Let us begin by noting that when the limit $1/\sigma \to 0$ is taken in the formal solution (1.35), one recovers the well known GR solution (1.5), namely

$$e^{-\lambda} = 1 - \frac{k^2}{r}\int_0^r \rho x^2\,dx\,, \tag{1.45}$$

and the GR limit does not seem an issue anymore. However, the problem is that the naive expression (1.35) is not a true solution at all. This can be seen by making use of Eq. (1.37) in the expression (1.31) to obtain

$$\tilde{\rho} = \rho - \frac{\rho}{\sigma}\,(\rho + 3\,p) + \frac{1}{8\,\pi}\,(2\,G^2_{\ 2} + G^1_{\ 1}) - 3\,p\,. \tag{1.46}$$

Equation (1.35) can then be written as

$$e^{-\lambda} = 1 - \frac{k^2}{r}\int_0^r \rho x^2\,dx$$
$$- \frac{k^2}{r}\int_0^r \left[\frac{\rho}{\sigma}\,(\rho + 3\,p) - \frac{1}{8\pi}\,(2\,G^2_{\ 2} + G^1_{\ 1}) + 3\,p\right]x^2\,dx\,, \tag{1.47}$$

which depends itself on λ and λ' through $G^1_{\ 1}$ and $G^2_{\ 2}$. Equation (1.47) is an integro-differential equation for the geometrical function λ, something completely different from the GR case, and a direct consequence of the non-locality of the BW equations. A way to solve this issue will be explained next.

1.4   MGD Constraint on the Brane

So far we have two closely related problems, which make it difficult to study non-uniform BW stars, and which any BW observer has to face, namely: the possibility that GR cannot be recovered when the formal solution (1.41) is implemented and the appearance of an integro-differential equation when the standard solution (1.35) is enforced. A method already presented in [14] to overcome these two problems, which eventually leads to a constraint that reduces the degrees of freedom on the brane, is now going to be explained in detail.

First of all, let us write the differential equation (1.40) as

$$\lambda' e^{-\lambda} \left(\frac{\nu'}{2} + \frac{1}{r} \right) - e^{-\lambda} \left(\nu'' + \frac{\nu'^2}{2} + 2\frac{\nu'}{r} + \frac{1}{r^2} \right) + \frac{1}{r^2}$$

$$+ 3 k^2 p + \frac{k^2}{\sigma} \rho \, (\rho + 3\, p) = -\frac{\lambda' e^{-\lambda}}{r} + \frac{e^{-\lambda}}{r^2} - \frac{1}{r^2} + k^2 \rho \,, \qquad (1.48)$$

so that the right hand side contains the standard GR terms and the bulk effects which modify the GR equation are all grouped in the left hand side. It can be seen that not all of the latter terms are manifestly bulk contributions, since not all of them are proportional to σ^{-1}. Indeed only high energy terms are manifestly bulk contributions and appear in the second line, while terms whose source is the Weyl curvature are displayed in the first line. These non-local terms, which do not explicitly appear as bulk contributions, are easily mixed with GR terms, as is shown in the form (1.40). Hence the solution eventually found for λ by solving Eq. (1.40) will never have the expected form shown in (1.43), and it will not be possible to recover the GR solution by simply taking the limit $1/\sigma \to 0$.

Keeping in mind GR as the proper limit, and the fact that a BW observer should see a geometric deformation due to five-dimensional gravity, we expect to write the solution in the form (1.7), that is

$$e^{-\lambda} = \mu(r) + f(r) \qquad (1.49)$$

where the function $\mu = \mu(r)$ is the well known GR solution (1.5),

$$\mu \equiv 1 - \frac{2\,m(r)}{r} \,, \qquad (1.50)$$

which contains the usual GR mass function (1.6). The unknown *geometric deformation* $f = f(r)$ in (1.49) should have two sources: the extrinsic curvature and the five-dimensional Weyl curvature. Correspondingly, we expect

$$f = \frac{1}{\sigma}(high-energy\ terms) + (non-local\ terms) \,, \qquad (1.51)$$

where, according to (1.47), the non-local terms in (1.51) must be related to the anisotropy projected onto the brane.

Demanding that the form (1.49) satisfies (1.48), one finds that f is determined by the first order differential equation

$$f' + \left(\frac{\nu'' + \nu'^2/2 + 2\,\nu'/r + 2/r^2}{\nu'/2 + 2/r} \right) f = \frac{k^2 \rho \, (\rho + 3\, p)/\sigma - H(p, \rho, \nu)}{\nu'/2 + 2/r} \,,$$
$$\qquad (1.52)$$

where we introduced the function

$$H(p, \rho, \nu) \equiv 3 k^2 p - \left[\mu' \left(\frac{\nu'}{2} + \frac{1}{r} \right) + \mu \left(\nu'' + \frac{\nu'^2}{2} + \frac{2\nu'}{r} + \frac{1}{r^2} \right) - \frac{1}{r^2} \right].$$
(1.53)

The solution of Eq. (1.52) is given by

$$f = e^{-I} \int_0^r \frac{e^I}{\nu'/2 + 2/x} \left[\frac{k^2}{\sigma} \rho (\rho + 3 p) + H(p, \rho, \nu) \right] dx + \beta(\sigma) e^{-I}, \quad (1.54)$$

where the local (proportional to $1/\sigma$) and non-local (the term H) bulk effects are clearly shown inside the square brackets. The function I is given again by the expression (1.42) and β is an integration constant which must depend on the brane tension σ so that it vanishes in the GR limit $1/\sigma \to 0$.

It is also easy to see from Eqs. (1.35)–(1.37) evaluated for $\sigma^{-1} = 0$ (the GR limit) that the non-local function $H(p, \rho, \nu)$ can be written as

$$H(p, \rho, \nu) = 3 k^2 p - \left(2 G^2_{2} + G^1_{1} \right) |_{\sigma^{-1} = 0}, \quad (1.55)$$

which clearly corresponds to an anisotropic term. In the GR case with an isotropic perfect fluid as matter source, the function $H(p, \rho, \nu)$ vanishes as a consequence of the isotropy of the solution. However, in the BW, the isotropic condition $H(p, \rho, \nu) = 0$ will not be satisfied anymore in general. There is in fact no reason to believe that the modifications undergone by p, ρ and ν, due to the bulk effects on the brane, do not modify the isotropic condition and one expects that $H(p, \rho, \nu) \neq 0$. The solution for the geometric function λ is finally given by

$$e^{-\lambda} = \underbrace{1 - \frac{k^2}{r} \int_0^r \rho x^2 \, dx}_{\text{GR-solution}}$$

$$\underbrace{+ e^{-I} \int_0^r \frac{e^I}{\nu'/2 + 2/x} \left[H(p, \rho, \nu) + \frac{k^2}{\sigma} \rho (\rho + 3 p) \right] dx + \beta(\sigma) e^{-I}}_{\text{Geometric deformation}},$$

$$\equiv \mu(r) + f(r). \quad (1.56)$$

As we mentioned before, we will usually describe spherically symmetric compact objects by means of a perfect fluid. In that case, in the region inside the object, the condition $\beta(\sigma) = 0$ has to be imposed in order to avoid singular solutions at the center $r = 0$.

The function $H(p, \rho, \nu)$ encodes anisotropic effects due to the bulk gravity affecting p, ρ and ν. It is worth emphasising that the geometric deformation $f(r)$ shown in Eq. (1.56) indeed "distorts" the GR solution given in Eq. (1.50) and induces an anisotropy (1.34) whose explicit expression from Eq. (1.36) is found to be

$$\Pi = \mu \left(\frac{1}{r^2} + \frac{\nu'}{r} \right) - \frac{\mu}{4} \left(2\nu'' + \nu'^2 + 2\frac{\nu'}{r} \right) - \frac{\mu'}{4} \left(\nu' + \frac{2}{r} \right) - \frac{1}{r^2}$$
$$+ f \left(\frac{1}{r^2} + \frac{\nu'}{r} \right) - \frac{f}{4} \left(2\nu'' + \nu'^2 + 2\frac{\nu'}{r} \right) - \frac{f'}{4} \left(\nu' + \frac{2}{r} \right) . \quad (1.57)$$

The above displays clearly that the overall anisotropy contains two possible contributions, from μ and f, respectively.

Now, in order to recover the GR solution in the expression given by Eq. (1.56), the following condition must be satisfied

$$\lim_{\sigma^{-1} \to 0} \int_0^r \frac{e^{I(x)} H(p, \rho, \nu)}{\nu'/2 + 2/x} \, dx = 0 . \quad (1.58)$$

This can be interpreted as a "constraint" whose physical meaning is nothing but the necessary condition to recover GR in the limit of infinite brane tension. A crucial observation is now that, when a given (spherically symmetric and isotropic) perfect fluid solution in GR is considered as a candidate solution for the BW system of Eqs. (1.27)–(1.30) [or, equivalently, Eq. (1.30) along with Eqs. (1.35)–(1.37)], one identically obtains

$$H(p, \rho, \nu) = 0 , \quad (1.59)$$

therefore every isotropic perfect fluid solution in GR will generate a *minimal* deformation of the radial metric component (1.56), given by

$$f^*(r) = \frac{k^2}{\sigma} e^{-I(r)} \int_0^r \frac{e^{I(x)} \rho (\rho + 3p)}{\nu'/2 + 2/x} \, dx . \quad (1.60)$$

The function f^* in Eq. (1.60) is proportional to $1/\sigma$ and represents a minimal deformation in the sense that all sources of the deformation in (1.56) have been removed, except for the matter density and pressure, which will always be necessary for describing a realistic stellar distribution. Consequently, the minimal deformation f^* induces a minimal anisotropy proportional to $1/\sigma$ and given by

$$\Pi = f^* \left(\frac{1}{r^2} + \frac{\nu'}{r} - \frac{\nu''}{2} - \frac{\nu'^2}{4} - \frac{\nu'}{2r} \right) - \frac{(f^*)'}{4} \left(\nu' + \frac{2}{r} \right) . \quad (1.61)$$

1.5 Summary: MGD Procedure in the Brane World

We have seen that the constraint (1.58) explicitly ensures the GR limit through the solution (1.56), and is very useful for generating solutions which recover GR as a limit [14]. It is also clearly established from Eq. (1.55) that the constraint (1.59) represents a condition of isotropy in GR. Thus in the context of the BW, the constraint

(1.59) has a direct physical interpretation: *Any bulk corrections to p, ρ and ν will only induce anisotropic effects on the brane which vanish in the GR limit* $1/\sigma \to 0$ (see Fig. 1.4). We can therefore say that the constraint (1.59) represents a natural way to generalise isotropic perfect fluid solutions from GR to the BW.

> **We can now summarise the main steps one should follow in order to apply the MGD to BW configurations:**

- Step 1: Choose a known isotropic GR solution (p, ρ, ν) of the conservation equation $p' = -\nu'(\rho + p)/2$.
- Step 2: Impose the constraint $H(p, \rho, \nu) = 0$ to make sure we have a solution for the geometric function λ with the correct GR limit (see Fig. 1.5):

$$e^{-\lambda} = 1 - \frac{k^2}{r} \int_0^r \rho x^2 \, dx + \frac{k^2}{\sigma} e^{-I(r)} \int_0^r \frac{e^{I(x)} \rho (\rho + 3p)}{\nu'/2 + 2/x} \, dx .$$

- Step 3: Compute $\Pi \sim \mathcal{P}$ and $\mathcal{U}$ from the Eqs. (1.36) and (1.37).
- Step 4: Replace the condition of vanishing pressure at the surface with matching conditions between the interior and exterior geometries. These will relate the integration constants to the brane tension, $C \to C(\sigma)$, and allow to determine the BW effects on the pressure and density.

There is still something which we might be concerned about. We can see that Eq. (1.30) does not show any manifest bulk contribution, and one might think that the solution eventually found for p, ρ and ν were the same as in GR. Indeed the bulk contributions for p, ρ and ν appear throughout the matching conditions at the surface of the compact object, as we mentioned in Step 4 above. In order to explain this important issue in details, we first need to analyse the consequences of the matching conditions for the MGD.

1.6 Matching Conditions for Stellar Distributions

The typical systems we wish to study are compact objects whose interior geometry (for $0 \le r < R_s$) is different from the outer geometry (for $r > R_s$). The matching conditions at the stellar surface $r = R_s$ between these two geometries are of particular importance, since from the observational point of view one has only direct access to the outer geometry.

Fig. 1.4 When the
constraint $H(\rho, p, \nu) = 0$ is
imposed, the
extra-dimensional effects
$(\delta p, \delta \rho, \delta \nu)$ on the set
(p, ρ, ν) only produce the
minimal anisotropy
proportional to $1/\sigma$ on the
brane. This ensures the
correct low-energy (GR)
limit

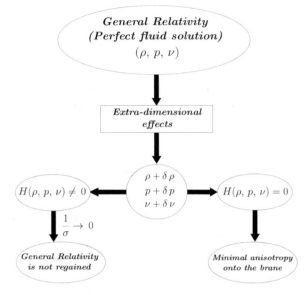

Fig. 1.5 MGD prescription
and GR limit for a compact
source

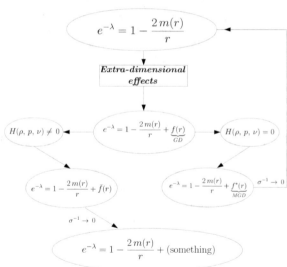

Let us assume that the interior geometry is given by a metric obtained by applying the MGD prescription, and reads

$$ds^2_{(-)} = e^{\nu_{(-)}} dt^2 - e^{\lambda_{(-)}} dr^2 - r^2 \left(d\theta^2 + \sin^2\theta d\phi^2 \right) , \qquad (1.62)$$

with

$$e^{-\lambda_{(-)}} = 1 - \frac{2\,\tilde{m}(r)}{r} , \qquad (1.63)$$

where the interior mass function is given by

$$\tilde{m}(r) = m(r) - \frac{r}{2} f^*(r) , \qquad (1.64)$$

with f^* the MGD given in Eq. (1.60). Since for a regular matter distribution we expect ρ and p are both positive, from Eq. (1.60) we can actually see that the geometric deformation

$$f^*(r) \geq 0 , \qquad (1.65)$$

if $\nu'_{(-)} > -4/r$. In this case, the effective interior mass (1.64) is always reduced by the extra-dimensional effects.

The space outside the compact source of radius R_s does not contain any matter source, so that $p = \rho = 0$ for $r > R_s$. The MGD metric (1.62), characterising the star interior, should therefore be matched with an exterior geometry sourced only by the Weyl fluid defined by $\mathcal{U}^{(+)}$ and $\mathcal{P}^{(+)}$ [44]. This can be generically written as

$$ds^2_{(+)} = e^{\nu_{(+)}} dt^2 - e^{\lambda_{(+)}} dr^2 - r^2 \left(d\theta^2 + \sin^2\theta d\phi^2 \right) , \qquad (1.66)$$

where the explicit form of the functions $\nu_{(+)}$ and $\lambda_{(+)}$ are obtained by solving the effective four-dimensional vacuum Einstein equations. The latter can be obtained from Eq. (1.14) by setting all sources to zero in the effective energy-momentum tensor, except for the Weyl contribution, namely

$$R_{\mu\nu} - \frac{1}{2} g_{\mu\nu} R = \mathcal{E}_{\mu\nu} . \qquad (1.67)$$

which imply

$$R = 0 . \qquad (1.68)$$

We recall that extra-dimensional effects are contained in the projected Weyl tensor $\mathcal{E}_{\mu\nu}$ and that only a few analytical solutions are known to date [45–48].

Continuity of the first fundamental form at the star surface defined by $r = R_s$ reads [35]

$$\left[ds^2 \right]_\Sigma = 0 , \qquad (1.69)$$

where $[F]_\Sigma \equiv F(r \to R_s^+) - F(r \to R_s^-) \equiv F_{R_s}^+ - F_{R_s}^-$, for any function $F = F(r)$, and yields

$$e^{\nu_{(-)}(R_s)} = e^{\nu_{(+)}(R_s)} \,, \tag{1.70}$$

and

$$1 - \frac{2M}{R_s} + f_{R_s}^* = e^{-\lambda_{(+)}(R_s)} \,, \tag{1.71}$$

where $M = m(R_s)$. Likewise, continuity of the second fundamental form at the star surface reads [35]

$$\left[G_{\mu\nu}\, r^\nu\right]_\Sigma = 0 \,, \tag{1.72}$$

where r^μ is a unit radial vector orthogonal to $r = R_s$. Using the effective Einstein field equations (1.14), we then find

$$\left[T_{\mu\nu}^{\text{eff}}\, r^\nu\right]_\Sigma = 0 \,, \tag{1.73}$$

which leads to

$$\left[p + \frac{1}{\sigma}\left(\frac{\rho^2}{2} + \rho\, p + \frac{2}{k^4}\mathcal{U}\right) + \frac{4\mathcal{P}}{k^4\,\sigma}\right]_\Sigma = 0 \,. \tag{1.74}$$

Since we assumed the star is surrounded by a Weyl fluid characterised by $\mathcal{U}^{(+)}$, $\mathcal{P}^{(+)}$, and $p = \rho = 0$ for $r > R_s$, this matching condition takes the form

$$p_{R_s} + \frac{1}{\sigma}\left(\frac{\rho_{R_s}^2}{2} + \rho_{R_s}\, p_{R_s} + \frac{2}{k^4}\mathcal{U}_{R_s}^{(-)}\right) + \frac{4\mathcal{P}_{R_s}^{(-)}}{k^4\,\sigma} = \frac{2\mathcal{U}_{R_s}^{(+)}}{k^4\,\sigma} + \frac{4\mathcal{P}_{R_s}^{(+)}}{k^4\,\sigma} \,, \tag{1.75}$$

where $p_{R_s} \equiv p_{R_s}^{(-)}$ and $\rho_{R_s} \equiv \rho_{R_s}^{(-)}$. Finally, by using Eqs. (1.37) and (1.61) in the condition (1.75), we obtain

$$p_{R_s} + \frac{f_{R_s}^*}{k^2}\left(\frac{\nu_{R_s}'}{R_s} + \frac{1}{R_s^2}\right) = \frac{2\mathcal{U}_{R_s}^{(+)}}{k^4\,\sigma} + \frac{4\mathcal{P}_{R_s}^{(+)}}{k^4\,\sigma} \,, \tag{1.76}$$

where $\nu_{R_s}' \equiv \partial_r \nu_{(-)}|_{r=R_s}$. Equations (1.70), (1.71) and (1.76) are the necessary and sufficient conditions for the matching of the interior MGD metric (1.62) with a spherically symmetric BW vacuum.

The matching condition (1.76) immediately yields an important result: if the exterior geometry is given by the Schwarzschild metric,

$$e^{\nu_{(+)}} = e^{-\lambda_{(+)}} = 1 - \frac{2\mathcal{M}}{r} \,, \tag{1.77}$$

one must have $\mathcal{U}^{(+)} = \mathcal{P}^{(+)} = 0$, which then leads to

$$p_{R_s} = -\frac{f_{R_s}^*}{k^2}\left(\frac{\nu_{R_s}'}{R_s} + \frac{1}{R_s^2}\right) .$$ (1.78)

Since we showed in Eq. (1.65) above that quite generally $f^* \geq 0$, an exterior vacuum can only be supported in the BW by exotic stellar matter, with negative pressure p_{R_s} at the surface, in agreement with the model discussed in Ref. [49]. We shall discuss this important point in more details in Chap. 2.

1.7 About Exact Solutions

We shall here construct a simple, mathematically consistent (albeit physically unsound, as we will detail below), solution in order to clarify how to apply the MGD procedure and obtain bulk corrections on the pressure p, the density ρ and the metric function ν (we recall that the metric function λ is determined by the MGD).

First of all, in order to evaluate the integral (1.42), and eventually obtain the minimal deformation (1.60), we need to consider a simple enough expression for the interior metric function $\nu_{(-)}$. Accordingly, we set

$$e^{\nu_{(-)}} = A\,r^4 ,$$ (1.79)

where A is a constant of suitable dimensions. Secondly, we enforce the continuity equation and the constraint $H(p, \rho, \nu) = 0$ to make sure that we will have a solution for the geometric function λ with the correct GR limit. Of course, the constraint (1.59) implies (1.58) and what it means is that eventual bulk corrections to p, ρ and ν will produce minimal anisotropic effects on the brane. A simple but unique solution to the continuity Eq. (1.30) and the constraint (1.59) which follows from (1.79) is given by

$$\begin{cases} \rho = \dfrac{B}{r^{4/3}} \\[3mm] k^2\,p = \dfrac{4}{r^2} - \dfrac{3\,k^2\,B}{r^{4/3}} , \end{cases}$$ (1.80)

where B is also a constant. We shall see later that A and B are promoted to functions of the brane tension σ by the matching conditions.

Using (1.79) and (1.80) in Eq. (1.56) we obtain

$$e^{-\lambda_{(-)}} = 1 - \frac{2\,M}{r}\left(\frac{r}{R_s}\right)^{5/3} + \frac{k^2\,B}{\sigma\,r^{4/3}}\left(\frac{18}{13} - \frac{12\,B}{7\,r^{4/3}}\right) ,$$ (1.81)

where

$$M = m(R_s) = k^2 \int_0^{R_s} \rho(r)\,r^2\,dr ,$$ (1.82)

and R_s is the radius of the distribution, as before. Using (1.36) and (1.37) we can obtain the Weyl functions

$$\mathcal{P}^{(-)} = \frac{B}{r^{8/3}} \left(\frac{4}{13\,k^2\,r^{2/3}} - \frac{4\,B}{17} \right)$$

$$\mathcal{U}^{(-)} = \frac{B}{r^{8/3}} \left(\frac{1}{13\,k^2\,r^{2/3}} - \frac{3\,B}{68} \right) .$$

(1.83)

It is important to remark right away that this interior solution is not regular at the origin $r = 0$, and cannot be taken as an acceptable candidate for describing a physical object. However, this singularity is just due to the choice of Eq. (1.80) and not the MGD prescription, which we intend to exemplify here.

The bulk contribution to p, ρ and ν can be found from the matching conditions. For instance, let us consider for the exterior region the Schwarzschild solution (1.77), where $\mathcal{M}$ is the ADM mass of the self-gravitating system and, correspondingly, we have

$$\mathcal{U}^{(+)} = \mathcal{P}^{(+)} = 0 ,$$

(1.84)

for $r > R_s$. The matching condition (1.69) yields

$$A\,R_s^4 = 1 - \frac{2\,\mathcal{M}}{R_s}$$

(1.85)

and

$$\mathcal{M} = M - \frac{k^2\,B}{\sigma\,R_s^{1/3}} \left(\frac{9}{13} - \frac{6\,B}{7\,R_s^{4/3}} \right) .$$

(1.86)

From (1.75) we then obtain

$$B = \frac{765 - 1326\,k^2\,\sigma\,R_s^2/4 + \sqrt{51\left(11475 + 3575\,k^2\,\sigma\,R_s^2 + 8619\,k^4\,\sigma^2\,R_s^4/2\right)}}{780\,k^2\,R_s^{2/3}} ,$$

(1.87)

which indeed depends on σ. Considering only terms up to the first order in $1/\sigma$, we have

$$B \simeq \frac{4}{3\,k^2\,R_s^{2/3}} + \frac{1960}{1989\,k^4\,\sigma\,R_s^{8/3}}$$

(1.88)

$$\mathcal{M} \simeq M - \frac{4}{\sigma\,R_s} \left(\frac{3}{13} - \frac{8}{21\,k^2} \right)$$

(1.89)

$$A \simeq \frac{1}{R_s^4} \left[1 - \frac{2\,M}{R_s} + \frac{8}{\sigma\,R_s^2} \left(\frac{3}{13} - \frac{8}{21\,k^2} \right) \right] ,$$

(1.90)

from which one obtains the interior quantities

$$p \simeq \frac{4}{k^2 \, r^2} - \frac{3}{k^2 \, R_s^{2/3} \, r^{4/3}} \left[\frac{4}{3} + \frac{1960}{1989 \, k^2 \, \sigma \, R_s} \right] \tag{1.91}$$

$$\rho \simeq \frac{4}{3 \, k^2 \, R_s^{2/3} \, r^{4/3}} \left[1 + \frac{490}{663 \, k^2 \, R_s} \right] \tag{1.92}$$

$$e^{\nu_{(-)}} \simeq \left[1 - \frac{2 \, M}{R_s} + \frac{8}{R_s^2} \left(\frac{3}{13} - \frac{8}{21 \, k^2} \right) \right] \left(\frac{r}{R_s} \right)^4 . \tag{1.93}$$

Finally, the outer quantities are given by

$$\mathcal{P}^{(+)} \simeq \frac{16}{k^4 \, R_s^{2/3} \, r^{8/3}} \left[\frac{3}{52 \, r^{2/3}} - \frac{4}{153 \, R_s^{2/3}} \right] \tag{1.94}$$

$$\mathcal{U}^{(+)} \simeq \frac{1}{3 \, k^4 \, R_s^{2/3} \, r^{8/3}} \left[\frac{1}{13 \, r^{2/3}} - \frac{1}{17 \, R_s^{2/3}} \right] \tag{1.95}$$

and

$$e^{-\lambda_{(+)}} \simeq 1 - \frac{2 \, M}{R_s} \left(\frac{r}{R_s} \right)^{2/3} + \frac{8}{\sigma \, R_s^{2/3} \, r^{2/3}} \left(\frac{3}{13 \, r^{2/3}} - \frac{8}{21 \, k^2 \, R_s^{2/3}} \right) , \tag{1.96}$$

which is not asymptotically flat.

The quantities (1.91)–(1.96) represent an exact analytical solution to the system (1.27)–(1.30) with the correct limit at low energies, but which contains a physically unacceptable singularity in the origin and is not asymptotically flat. In fact, this solution was constructed for the sole purpose of clarifying the method described in the previous sections and summarised in the algorithm of Sect. 1.5. In order to obtain a physically acceptable solution, with no spurious singularities, it is necessary to carry out a more careful analysis. This will be addressed in the next chapters.

Of course the fact that the constraint (1.59) is satisfied by every isotropic perfect fluid in GR does not mean that all of the GR solutions can be directly used to find an exact BW configuration by means of the MGD. In order to investigate whether a particular GR solution admits an exact BW extension, the first step is to analyse the temporal component of the metric. If ν is not simple enough, it will be hardly possible to find an analytic expression for the integral in (1.42). Even in the case where this expression is analytic, finding an exact expression for λ using Eq. (1.56) can be very difficult.

As a guide to find exact solutions, for the metric function ν, one can consider the more general expression

$$e^\nu = A \left(1 + C \, r^m \right)^n \tag{1.97}$$

which gives an analytic expression for the integral (1.42). This expression can then be used in (1.56), and a complicated integral equation for ρ when (1.97) is used in (1.30) and (1.59). It is difficult to figure out appropriate values for the set of constants $\{A, C, m, n\}$ leading to exact expressions for p and ρ, and even more so for the searching of exact and physically acceptable solutions. Nevertheless, exact solutions for $\{p(r), \rho(r), \nu(r)\}$ where found in Ref. [14] and an exact solution for the complete system $\{p(r), \rho(r), \nu(r), \lambda(r), \mathcal{U}(r), \mathcal{P}(r)\}$ was reported in Ref. [43]. It is worth noticing that in the case of a uniform distribution, the system is *closed*, thus it is not necessary to impose any additional restriction, except from the constraint (1.59), which will produce a non-linear differential equation for the geometric function ν.

References

1. C.M. Will, The confrontation between GR and experiment. Living Rev. Rel **9** (2006)
2. C.F. Sopuerta, Probing the strong gravity regime with eLISA: progress on EMRIs, arXiv:1210.0156 [gr-qc]
3. I. Ben-Dayan, M. Gasperini, G. Marozzi, F. Nugier, G. Veneziano, Average and dispersion of the luminosity-redshift relation in the concordance model. JCAP **1306**, 002 (2013)
4. O. Umeh, C. Clarkson, R. Maartens, Nonlinear general relativistic corrections to redshift space distortions, gravitational lensing magnification and cosmological distances. Class. Quant. Grav. **31**, 202001 (2014)
5. P.A.R. Ade et al., (Planck Collaboration): Planck 2013 results. I. Overview of products and scientific results. Astron. Astrophys. **571** A1 (2014)
6. C.L. Bennett et al., (WMAP Collaboration): nine-year Wilkinson Microwave Anisotropy Probe (WMAP) observations: final maps and results. Astrophys. J. Suppl. **208**, 20 (2013)
7. K. Akiyama et al., (EHT collaboration): Event horizon telescope. Astrophys. J. **875** (2-19) L1
8. K. Akiyama et al., (EHT collaboration): Event horizon telescope. Astrophys. J. **875** (2-19) L4
9. B.P. Abbott et al., (LIGO and VIRGO collaboration). Phys. Rev. Lett. **116**, 061102 (2016)
10. B.P. Abbott et al., (LIGO and VIRGO collaboration). Phys. Rev. Lett. **119**, 141101 (2017)
11. B.P. Abbott et al., (LIGO and VIRGO collaboration). J. Astrophys. **851**, L35 (2017)
12. M. Aguilar et al., (AMS Collaboration): First result from the alpha magnetic spectrometer on the international space station: precision measurement of the positron fraction in primary cosmic rays of 0.5–350 GeV. Phys. Rev. Lett. **110** 141102 (2013)
13. R. Agnese et al., (CDMS Collaboration): Silicon detector dark matter results from the final exposure of CDMS II. Phys. Rev. Lett. **111**(25), 251301 (2013)
14. J. Ovalle, Searching exact solutions for compact stars in braneworld: a conjecture. Mod. Phys. Lett. A **23**, 3247 (2008)
15. J. Ovalle, Braneworld stars: anisotropy minimally projected onto the brane, in *Gravitation and Astrophysics (ICGA9)*, ed. by J. Luo (World Scientific, Singapore, 2010), pp. 173–182
16. J. Ovalle, Effects of density gradients on braneworld stars, in *Proceedings of the Twelfth Marcel Grossmann Meeting on GR*, ed. by T. Damour, R. T. Jantzen, R. Ruffini (World Scientific, Singapore, 2012), pp. 2243–2245. ISBN 978-981-4374-51-4
17. R. Casadio, J. Ovalle, Brane-world stars and (microscopic) black holes. Phys. Lett. B **715**, 251 (2012)
18. J. Ovalle, F. Linares, Tolman IV solution in the Randall-Sundrum braneworld. Phys. Rev. D **88**(2013), 104026 (2013)
19. J. Ovalle, F. Linares, A. Pasqua, A. Sotomayor, The role of exterior Weyl fluids on compact stellar structures in Randall-Sundrum gravity. Class. Quant. Grav. **30**, 175019 (2013)
20. R. Casadio, J. Ovalle, Brane-world stars from minimal geometric deformation, and black holes. Gen. Relat. Grav. **46**, 1669 (2014)
21. R. Casadio, J. Ovalle, R. da Rocha, Black strings from minimal geometric deformation in a variable tension brane-world. Class. Quant. Grav. **30**, 175019 (2014)
22. R. Casadio, J. Ovalle, R. da Rocha, Classical tests of GR: brane-world sun from minimal geometric deformation. Europhys. Lett. **110**, 40003 (2015)

23. R. Casadio, J. Ovalle, R. da Rocha, The minimal geometric deformation approach extended. Class. Quant. Grav. **32**, 215020 (2015)
24. J. Ovalle, Extending the geometric deformation: new black hole solutions, in *International Journal of Modern Physics: Conference Series*, vol. 41 (2016), p. 1660132
25. L. Randall, R. Sundrum, A large mass hierarchy from a small extra dimension. Phys. Rev. Lett. **83**, 3370 (1999)
26. L. Randall, R. Sundrum, An alternative to compactification. Phys. Rev. Lett. **83**, 4690 (1999)
27. N. Arkani-Hamed, S. Dimopoulos, G. Dvali, The hierarchy problem and new dimensions at a millimeter. Phys. Lett. B **429**, 263 (1998)
28. I. Antoniadis, N. Arkani-Hamed, S. Dimopoulos, G. Dvali, New dimensions at a millimeter to a fermi and superstrings at a TeV. Phys. Lett. B **436**, 257 (1998)
29. R. Maartens, Brane world gravity. Living Rev. Rel. **7**, 7 (2004)
30. R. Maartens, K. Koyama, Brane-world gravity. Living Rev. Rel. **13**, 5 (2010)
31. G. Aad et al., (ATLAS Collaboration): Observation of a new particle in the search for the Standard Model Higgs boson with the ATLAS detector at the LHC. Phys. Lett. B **716**, 1 (2012)
32. S. Chatrchyan et al., (CMS Collaboration): Search for narrow resonances and quantum black holes in inclusive and b-tagged dijet mass spectra from pp collisions at $\sqrt{s} = 7$ TeV. JHEP **1301**, 013 (2013)
33. L.A. Gergely, T. Harko, M. Dwornik, G. Kupi, Z. Keresztes, Galactic rotation curves in brane world models. Mon. Not. Roy. Astron. Soc. **415**, 3275 (2011)
34. T. Shiromizu, K.I. Maeda, M. Sasaki, The Einstein equation on the 3-brane world. Phys. Rev. D **62**, 024012 (2000)
35. W. Israel, Singular hypersurfaces and thin shells in GR. Nuovo Cim. B **44**, 1 (1966); Nuovo Cim. B **48**, 463 (1966)
36. N. Arkani-Hamed, M. Schmaltz, Hierarchies without symmetries from extra dimensions. Phys. Rev. D **61**, 033005 (2000)
37. R. Casadio, L. Mazzacurati, Bulk shape of brane world black holes. Mod. Phys. Lett. A **18**, 651 (2003). R. Casadio, O. Micu, Exploring the bulk of tidal charged micro-black holes. Phys. Rev. D **81**, 104024 (2010)
38. R. da Rocha, J.M. Hoff da Silva, Black string corrections in variable tension braneworld scenarios. Phys. Rev. D **85**, 046009 (2012)
39. J. Campbell, *A Course of Differential Geometry* (Clarendon, Oxford, 1926); L. Magaard, Ph.D. thesis, University of Kiel, 1963
40. M.D. Maia, Hypersurfaces of five dimensional space-times, arXiv:gr-qc/9512002v2
41. K. Koyama, R. Maartens, Structure formation in the DGP cosmological model. JCAP **01**, 016 (2006)
42. A. Viznyuk, Y. Shtanov, Spherically symmetric problem on the brane and galactic rotation curves. Phys. Rev. D **76**, 064009 (2007)
43. J. Ovalle, Non-uniform braneworld stars: an exact solution. Int. J. Mod. Phys. D **18**, 837 (2009)
44. C. Germani, R. Maartens, Stars in the braneworld. Phys. Rev. D **64**, 124010 (2001)
45. N. Dadhich, R. Maartens, P. Papadopoulos, V. Rezania, Black holes on the brane. Phys. Lett. B **487**, 1 (2000)
46. R. Casadio, A. Fabbri, L. Mazzacurati, New black holes in the brane world? Phys. Rev. D **65**, 084040 (2002)
47. P. Figueras, T. Wiseman, Gravity and large black holes in Randall-Sundrum II braneworlds. Phys. Rev. Lett. **107**, 081101 (2011)
48. J. Ovalle, F. Linares, Tolman IV solution in the Randall-Sundrum braneworld. Phys. Rev. D **88**, 104026 (2013)
49. L.A. Gergely, Black holes and dark energy from gravitational collapse on the brane. JCAP **02**, 027 (2007)

Chapter 2
Stellar Distributions

It is notorious that exact solutions representing the interior of a compact object are hard to find in GR [1], even for the simple case of a static perfect fluid. Only very few of them are then also of physical interest [2]. Hence it is not surprising that finding exact solutions in the BW, where new terms arise on the brane because gravity propagates in the bulk, becomes an extremely complicated task. The reason is that the (four-dimensional) non-locality makes the BW equations an open system. As we have seen in the previous chapter, the non-locality derives from the projection of the bulk Weyl tensor onto the brane, and this leads to a very complicated system of equations which are particularly difficult to study for non-uniform matter distributions. Indeed, there is no well-established criteria about what restrictions should be considered in the BW equations in order to obtain a closed system. Although there are exceptions, for instance, a closed system of equations for static spherically symmetric system with $\mathcal{P} = 0$ was studied in Ref. [3], a better understanding of the bulk geometry and how our four-dimensional spacetime is embedded in it is necessary in order to solve this issue.

Since the complete five-dimensional problems (bulk plus brane) remain unsolved, finding exact, and possibly physically acceptable solutions to the effective four-dimensional Einstein field equations (1.14), appears like a good guide to clarify aspects of the five-dimensional gravity and the way our observed universe is embedded in it. Once a solution is found, we could use the Campbell-Magaard theorems [4, 5] to extend the BW solution into the bulk, at least locally (See Ref. [6, 7] for the Campbell-Magaard theorems in GR). First of all we need to start with a rigorous and systematic study of the effective field Eq. (1.14). A study that, among other things, should clarify the role of five-dimensional effects onto the brane and that can consolidate a general methodology based on the critical experimental requirement that GR be recovered at low energies. In this respect, the RS model is characterised

© The Author(s), under exclusive license to Springer Nature Switzerland AG 2020
J. Ovalle and R. Casadio, *Beyond Einstein Gravity*, SpringerBriefs in Physics,
https://doi.org/10.1007/978-3-030-39493-6_2

Fig. 2.1 Schematic picture
of a BW star: the interior is
characterised by the density
ρ, pressure p and radius R_s,
the exterior geometry by the
ADM mass $\mathcal{M}$ and the tidal
charge q. The brane has
tension σ, fine-tuned with
the bulk cosmological
constant, so that the brane
cosmological constant
$\Lambda = 0$

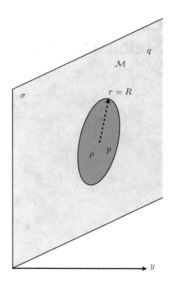

by the brane tension, which can be used as a parameter to control this limit [8] by
means of the MGD approach introduced in Chap. 1.

In this chapter, five different GR solutions representing compact objects will be
used to generate BW stellar interiors by applying the MGD prescription. Three of
these solutions represent exact and physically acceptable solutions to the system of
spherically symmetric BW equations (1.27)–(1.30). Moreover, the first four solutions
will be obtained by assuming that the matter inside the star is represented by a regular
fluid with positive pressure everywhere. The outer metric must therefore differ from
the vacuum Schwarzschild solution, as we have shown that follows in general from
the matching conditions at the star surface in Sect. 1.6. The only modified outer metric
we consider here is given by the DMPR solution [9] characterised by the tidal charge
q (see Fig. 2.1 and Sect. 2.1.1.2 below for the details), although several alternatives
exist in the literature [10–13]. We shall finally show that the price to pay in order to
preserve the Schwarzschild vacuum outside the star is to allow for a layer of solid
crust which we will analyse in details in the fifth solution.

2.1 The Heintzmann Solution

The MGD approach developed in the previous chapter ensures GR is recovered in
the limit $1/\sigma \to 0$, but it does not automatically make the solution thus found of
physical relevance. By this we mean that the set of variables which describe the
system, such as the density, pressure and asymptotic form of the metric at infinity,
satisfies the conditions one expect for compact objects made of regular matter. We

will see that the matching conditions with the exterior geometry play a main role in this important issue by constructing an explicit solution.

Let us consider in the interior the metric component $\nu_{(-)}$, pressure p and density ρ for a perfect fluid found by Heintzmann [14], that is

$$e^{\nu_{(-)}} = A \left(1 + C r^2\right)^3 \tag{2.1}$$

$$k^2 p = \frac{9 C \left(1 - C r^2\right)}{2 \left(1 + C r^2\right)^2} \tag{2.2}$$

$$k^2 \rho = \frac{3 C \left(3 + C r^2\right)}{2 \left(1 + C r^2\right)^2} . \tag{2.3}$$

Like for any perfect fluid in GR, these quantities satisfy the constrain (1.59). In GR, A and C above are arbitrary constants, but they become functions of the brane tension σ in the BW.

The solution for the geometric function $\lambda_{(-)}$ is obtained using (2.1)–(2.2) in (1.56), which leads to the usual form (1.63), with the interior mass function $\tilde{m}$ given by the expression in Eq. (1.64) and the minimal geometric deformation reads

$$f^* = \frac{g(r)}{\sigma k^2 r \left(1 + C r^2\right)(2 + 5 C r^2)^{11/10}} \tag{2.4}$$

where

$$g = 9 C^2 \int_0^r \frac{r^2 \left(2 + 5 C r^2\right)^{\frac{1}{10}} \left(9 - 3 C r^2 - 2 C^2 r^4\right)}{\left(1 + C r^2\right)^2} \, dr . \tag{2.5}$$

Since the standard GR mass function (1.6) is now given by

$$m = \frac{3 k^2 C r^3}{4 \left(1 + C r^2\right)} , \tag{2.6}$$

the total GR mass of the source of radius R_s is given by

$$M \equiv m(R_s) = \frac{3 k^2 C R_s^3}{4 \left(1 + C R_s^2\right)} . \tag{2.7}$$

A numerical analysis shows that $g(r) > 0$ for all $r \leq R_s$. Hence, we can see from (1.64) that the effective interior mass function $\tilde{m}$ is reduced by five-dimensional effects, that is $\tilde{m} < m$.

The interior Weyl functions are next obtained from (1.36) and (1.37) and read

$$\frac{\mathcal{P}^{(-)}}{k^2 \sigma} = \frac{1}{4 r^3 \left(1 + C r^2\right)^4 \left(2 + 5 C r^2\right)^{\frac{21}{10}}}$$
$$\times \left[2 \left(1 + C r^2\right)^2 \left(1 + 7 C r^2 + 14 C^2 r^4\right) g(r)\right.$$
$$-3 C^2 r^3 \left(2 + 5 C r^2\right)^{\frac{1}{10}}$$
$$\left. \times \left(18 + 111 C r^2 + 137 C^2 r^4 - 86 C^3 r^6 - 40 C^4 r^8\right)\right] , \quad (2.8)$$

and

$$\frac{\mathcal{U}^{(-)}}{k^2 \sigma} = \frac{C}{16 r \left(1 + C r^2\right)^4 \left(2 + 5 C r^2\right)^{\frac{21}{10}}} \cdot$$
$$\times \left[8 \left(1 + C r^2\right)^2 \left(5 + 7 C r^2\right) g(r)\right.$$
$$-3 C r \left(2 + 5 C r^2\right)^{\frac{1}{10}}$$
$$\left. \times \left(180 + 660 C r^2 + 509 C^2 r^4 - 62 C^3 r^6 - 55 C^4 r^8\right)\right] . \quad (2.9)$$

Of course, the completely explicit expressions of $\lambda, \mathcal{U}$ and $\mathcal{P}$ are not available because of the integral (2.5) that defines $g = g(r; C)$. However, this feature does not prevent the construction of a physically relevant exact solution for the pressure and density by means of the matching conditions at the star surface, where the assumption of vanishing pressure will be dropped [15, 16].

2.1.1   Matching Conditions

In GR, Birkhoff's theorem ensures that the vacuum outside a spherically symmetric source has the unique Schwarzschild geometry. This situation changes dramatically in the five-dimensional BW scenario, where high-energy corrections and the presence of the Weyl stresses imply that the exterior solution for a spherically symmetric distribution is no longer unique. This means that there are many possible candidates to describe the outer geometry.

2.1.1.1   Schwarzschild Exterior

The simplest choice of outer vacuum geometry is still given by the Schwarzschild metric (1.77). The continuity conditions (1.70) and (1.71) at the surface $r = R_s$ yield

$$A = \left(1 - \frac{2\mathcal{M}}{R_s}\right) \frac{1}{\left(1 + C R_s^2\right)^3} \qquad (2.10)$$

and

$$\mathcal{M} = M - \frac{g(R_s; C)}{2\,\sigma\,k^2\,(1 + C\,R_s^2)(2 + 5\,C\,R_s^2)^{11/10}}\,, \tag{2.11}$$

where we again remark that the function g in Eq. (2.5) depends on the constant C. From the second matching condition (1.75), we find that C must satisfy

$$\sigma\,k^2\,p(R_s) + \frac{g(R_s; C)\,\left(1 + 7\,C\,R_s^2\right)}{R_s^3\,\left(1 + C\,R_s^2\right)^2\,\left(2 + 5\,C\,R_s^2\right)^{\frac{11}{10}}} = 0\,. \tag{2.12}$$

Upon replacing the expression (2.2), the above equation becomes

$$\sigma\,\frac{9\,C\,\left(1 - C\,R_s^2\right)}{2\left(1 + C\,R_s^2\right)^2} + \frac{g(R_s; C)\,\left(1 + 7\,C\,R_s^2\right)}{R_s^3\,\left(1 + C\,R_s R^2\right)^2\,\left(2 + 5\,C\,R_s^2\right)^{\frac{11}{10}}} = 0\,, \tag{2.13}$$

which shows that C cannot take the simple GR value[1]

$$C_0 = \frac{1}{R_s^2}\,, \tag{2.14}$$

which follows from the condition $p(R_s) = 0$ in (2.2). We must conclude that the GR value C_0 is modified by bulk effects, and it should therefore be possible to write

$$C = C_0 + \delta C(\sigma)\,, \tag{2.15}$$

where $\delta C(\sigma)$ represents the "bulk perturbation" which vanishes for $1/\sigma \to 0$. In fact, at first order in σ^{-1}, one finds

$$\delta C(\sigma) \simeq \frac{16\,g(R_s; C_0 = R_s^{-2})}{9\,\sigma\,k^2\,7^{11/10}\,R_s^3}\,. \tag{2.16}$$

The pressure can now be found by expanding p around C_0,

$$p(C_0 + \delta C) \simeq p(C_0) + \delta C\,\left.\frac{dp}{dC}\right|_{C=C_0}\,, \tag{2.17}$$

which yields

$$k^2\,p = \frac{9\,R_s\,\left(R_s^2 - r^2\right)}{2\,\left(R_s^2 + r^2\right)^2} + \underbrace{\frac{9\,R_s^4\,\left(R_s^2 - 3\,r^2\right)}{2\,\left(R_s^2 + r^2\right)^3}\,\delta C(\sigma)}_{\delta p(\sigma)}\,. \tag{2.18}$$

---

[1]A numerical analysis shows that $g(R_s; C) \neq 0$ for a wide range of values of R_s.

The density can be found in the same way and reads

$$k^2 \rho(r) = \frac{3\,R_\mathrm{s}\left(3\,R_\mathrm{s}^2 + r^2\right)}{2\left(R_\mathrm{s}^2 + r^2\right)^2} + \underbrace{\frac{3\,R_\mathrm{s}^4\left(3\,R_\mathrm{s}^2 - r^2\right)}{2\left(R_\mathrm{s}^2 + r^2\right)^3}\,\delta C(\sigma)}_{\delta\rho(\sigma)}, \tag{2.19}$$

where $\delta p(\sigma)$ and $\delta\rho(\sigma)$ represent the bulk corrections on p and ρ respectively. In particular, at the surface and for any arbitrary radius R_s, we always have

$$k^2\,p(R_\mathrm{s}) = -\frac{9}{8}\,\delta C(\sigma) < 0\,, \tag{2.20}$$

which implies that the Schwarzschild exterior geometry and the interior solution found here are incompatible with a regular fluid, for which the pressure should be non-negative everywhere. If we want to model the matter inside the star with a regular fluid, we must therefore consider a different case.

2.1.1.2   Tidally Charged Exterior

As we mentioned, the presence of Weyl stresses implies that the exterior geometry for a spherically symmetric distribution is not necessarily the Schwarzschild metric. There are many possible solutions to the effective four-dimensional vacuum Einstein equations (1.14), although only a few analytical solutions are known.

One of these solutions is the DMPR, or tidally charged metric of Ref. [9], given by

$$e^{\nu_{(+)}} = e^{-\lambda_{(+)}} = 1 - \frac{2\,\mathcal{M}}{r} - \frac{q}{r^2}\,, \tag{2.21}$$

corresponding to the Weyl functions

$$\mathcal{U}^{(+)} = -\frac{\mathcal{P}^{(+)}}{2} = \frac{\sigma\,k^2\,q}{6\,r^4}\,. \tag{2.22}$$

This solution was extensively studied in Refs. [17, 18], and will be the exterior solution surrounding the BW star in this section.

The matching conditions given by Eqs. (1.70), (1.71) now lead to

$$A\left(1 + C\,R_\mathrm{s}^2\right)^3 = 1 - \frac{2\,\mathcal{M}}{R_\mathrm{s}} - \frac{q}{R_\mathrm{s}^2}\,, \tag{2.23}$$

and

$$\mathcal{M} = M - \frac{g(R_\mathrm{s}; C)}{2\,\sigma\,k^2\left(1 + C\,R_\mathrm{s}^2\right)\left(2 + 5\,C\,R_\mathrm{s}^2\right)^{11/10}} + \frac{q}{2\,R_\mathrm{s}}\,, \tag{2.24}$$

whereas Eq. (1.75) yields

$$q = -\frac{9 C R_s^4 \left(1 - C R_s^2\right)}{2 \left(1 + C R_s^2\right)^2} - \frac{R_s \left(1 + 7 C R_s^2\right) g(R_s; C)}{\sigma k^2 \left(1 + C R_s^2\right)^2 \left(2 + 5 C R_s^2\right)^{\frac{11}{10}}} . \tag{2.25}$$

As in the previous case, the constants C and A have well defined values in GR, denoted by A_0 and C_0, which are given respectively by the condition $p(R_s) = 0$ in (2.2) and evaluating (2.23) in the limit $1/\sigma \to 0$,

$$A_0 \left(1 + C_0 R_s^2\right)^3 = 1 - \frac{2 M}{R_s} . \tag{2.26}$$

On comparing (2.23) with (2.26), it is clear that the GR values are modified by bulk effects and the BW values depend on the tension σ, that is

$$C = C_0 + \delta C(\sigma) , \quad A = A_0 + \delta A(\sigma) . \tag{2.27}$$

In particular, from (2.27) we can write (2.23) as

$$(A_0 + \delta A) \left[1 + (C_0 + \delta C) R_s^2\right]^3 = 1 - \frac{2 M}{R_s} + \frac{q}{R_s^2} . \tag{2.28}$$

Upon replacing Eqs. (2.24) and (2.25) into this equation and expanding to linear order in $1/\sigma$, we find

$$C \simeq C_0 + \frac{8}{3 R_s^2} \left[\frac{g(R_s; C_0)}{2 \sigma k^2 7^{11/10} R_s} - 8 \delta A\right] . \tag{2.29}$$

The constant $C = C(\sigma)$ can then be determined by fixing $\delta A = \delta A(\sigma)$, and we can do so in such a way to ensure that the pressure p and density ρ computed from (2.2) and (2.3) are physically acceptable. Figure 2.2 compares the behaviour of the pressure in the GR and BW cases. The five-dimensional gravity effects reduce the pressure deep inside the distribution. However the situation changes for the outer layers, where the matching conditions lead to $p > 0$ at the surface.

It is important to remark that exact solutions for the pressure and density were found which represent a physically acceptable configuration. In fact, all quantities are regular at the origin, pressure and density are everywhere positive and decrease monotonically towards the surface, mass and radius are well-defined, etc. We found that this solution is incompatible with the Schwarzschild metric outside. Using the Reissner-Nördstrom-like metric given in Ref. [9], the effects of five-dimensional gravity on pressure and density compatible with the matching conditions were shown to reduce the pressure deep inside the distribution, but yield a positive pressure at the surface, in order to balance the outer pressure from the Weyl fluid.

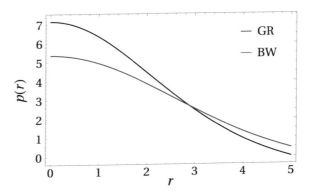

Fig. 2.2 Heintzmann BW solution. Qualitative comparison of the pressure $[p \times 10^3]$ in GR $[p(R_s) = 0]$ and in the BW $[p(R_s) \neq 0]$ with $R_s = 5$

2.2 The Interior Schwarzschild Solution

We next proceed to consider the interior Schwarzschild solution of GR. We recall that this metric generated by a homogeneous density is known analytically and matches smoothly with the Schwarzschild exterior metric [19]. Therefore the study of this interior solution in the BW represents a scenario of great interest. Indeed, a first BW generalisation of the Schwarzschild interior solution was obtained in the pioneering work of Germani and Maartens [20] by including only high-energy terms. This left out of the analysis the possible effects of the Weyl functions inside uniform distributions, as was eventually elucidated in Ref. [21]. In this section, we will present another extension of the Schwarzschild interior metric to the BW, in which both local terms (high-energy corrections) and non-local terms (from the bulk Weyl curvature) are considered.

Let us begin by writing the general Schwarzschild interior solution in GR as

$$e^{\nu_s} = \left(A - B\sqrt{1 - C r^2}\right)^2 \tag{2.30}$$

$$e^{-\lambda_s} = 1 - C r^2 , \tag{2.31}$$

where the constants A, B and C are related to the density

$$\rho = \frac{3 C}{k^2} , \tag{2.32}$$

and pressure

$$p = \frac{\rho}{3}\left(\frac{3 B \sqrt{1 - C r^2} - A}{A - B \sqrt{1 - C r^2}}\right) . \tag{2.33}$$

In GR, the matching conditions with the unique Schwarzschild exterior metric at the surface $r = R_s$ imply $B = 1/2$, but it is not necessarily so in the BW, where there are many possible exterior solutions. Moreover, all three constants A, B and C will be promoted to functions of the brane tension σ by the matching conditions with a convenient exterior solution.

We also recall that, the condition $p(R_s) = 0$ in GR[2] implies

$$A = 3 B \sqrt{1 - C R_s^2}\,, \tag{2.34}$$

yielding

$$e^{\nu(-)} = B^2 \left(3 \sqrt{1 - C R_s^2} - \sqrt{1 - C r^2} \right)^2 \tag{2.35}$$

and

$$p = \rho \left(\frac{\sqrt{1 - C r^2} - \sqrt{1 - C R_s^2}}{3 \sqrt{1 - C R_s^2} - \sqrt{1 - C r^2}} \right). \tag{2.36}$$

The quantities in Eqs. (2.32), (2.35) and (2.36) automatically satisfy the MGD constraint (1.59), and the bulk effects on them can therefore be found by means of the matching conditions only. On the other hand, the BW version for the radial metric component (2.31) is obtained using (2.32), (2.35) and (2.36) in (1.56), leading to the same expression (1.63), that is

$$e^{-\lambda(-)} = 1 - \frac{2\,\tilde{m}}{r}\,. \tag{2.37}$$

The interior mass function is here given by

$$\tilde{m} = m(r) - \frac{9\,r\,C^2\,g(r)}{\sigma\,k^2}\,, \tag{2.38}$$

where the GR mass function (1.6) is given by

$$m = \frac{C\,r^3}{2}\,, \tag{2.39}$$

so that the total GR mass is

$$M \equiv m(R_s) = \frac{C\,R_s^3}{2}\,. \tag{2.40}$$

The function

$$g = e^{-I(r)} \int_0^r \frac{e^{I(x)}\left(1 - C\,x^2\right) x\,dx}{3\,C\,x^2 - 2 + 2\,\gamma\,\sqrt{1 - C\,x^2}}\,, \tag{2.41}$$

with

---

[2]We already saw that this condition can be dropped for BW stars [15, 16].

$$\gamma \equiv 3\sqrt{1 - C\,R_s^2} \tag{2.42}$$

and I is given by the integral in Eq. (1.42). For the present case, it is useful to notice that this integral can be written as

$$I = \nu + 2\ln\left(\frac{\nu'}{2} + \frac{2}{r}\right) + 12\int_0^r \frac{dx}{x\,(x\,\nu' + 4)}$$

$$= \nu + 2\ln\left(\frac{\nu'}{2} + \frac{2}{r}\right) + \Phi, \tag{2.43}$$

where

$$e^\Phi = \left[\frac{\left(\sqrt{3 + \gamma^2} - \gamma\right) + 3\sqrt{1 - C\,r^2}}{\left(\sqrt{3 + \gamma^2} + \gamma\right) - 3\sqrt{1 - C\,r^2}}\right]^{\frac{\gamma}{2\sqrt{3 + \gamma^2}}}$$

$$\times \frac{\left(\sqrt{C}\,r\right)^3}{\sqrt{3\,C\,r^2 - 2 + 2\gamma\sqrt{1 - C\,r^2}}}. \tag{2.44}$$

2.2.1 The Interior Weyl Fluid

From Eqs. (1.36) and (1.37), the interior Weyl functions are given by

$$\frac{\mathcal{P}^{(-)}}{k^2\,\sigma} = \frac{\left[Cr^2\left(7 - 3\gamma\sqrt{1 - Cr^2}\right) + 3\gamma\sqrt{1 - Cr^2} - 6C^2r^4 - 3\right]\tilde{m}(r)}{6r^3\left(1 - Cr^2\right)^{3/2}\left(\sqrt{1 - Cr^2} - \gamma\right)}$$

$$+ \frac{\left(Cr^2 - 1\right)\left(\gamma\sqrt{1 - Cr^2} + 2Cr^2 - 1\right)\tilde{m}'(r) + C^2r^4}{6r^2\left(1 - Cr^2\right)^{3/2}\left(\sqrt{1 - Cr^2} - \gamma\right)} \tag{2.45}$$

and

$$\frac{\mathcal{U}^{(-)}}{k^2\,\sigma} = \frac{C\left[Cr^2\left(3\gamma\sqrt{1 - Cr^2} - 14\right) - 3\gamma\sqrt{1 - Cr^2} + 9C^2r^4 + 3\right]}{6\left(1 - Cr^2\right)^{3/2}\left(\sqrt{1 - Cr^2} - \gamma\right)}$$

$$+ \frac{3C^2\left(\gamma\sqrt{1 - Cr^2} - 3Cr^2 + 3\right)}{4k^2\sigma\sqrt{1 - cr^2}\left(\sqrt{1 - Cr^2} - \gamma\right)} - \frac{C\left(3Cr^2 - 5\right)\tilde{m}(r)}{3r\left(1 - Cr^2\right)^{3/2}\left(\sqrt{1 - Cr^2} - \gamma\right)}$$

$$- \frac{\left(\gamma\sqrt{1 - Cr^2} + 2Cr^2 - 1\right)\tilde{m}'(r)}{3r^2\left(\gamma\sqrt{1 - Cr^2} + Cr^2 - 1\right)}. \tag{2.46}$$

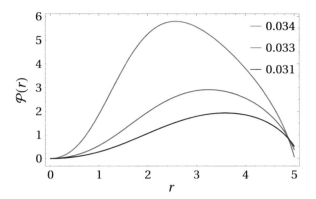

Fig. 2.3 Schwarzschild solution. Weyl function $[\mathcal{P}^{(-)} \times 10^4]$ for three different values of C with $R_s = 5$

Although both Weyl functions have complicated expressions, it is possible to identify some general features.

For instance, we have

$$\frac{\mathcal{U}^{(-)}(0)}{k^2 \sigma} = -\frac{3}{4} C^2 \frac{\gamma + 1}{\gamma - 1} , \qquad (2.47)$$

which shows a divergence in the origin when $\gamma = 1$. On using Eqs. (2.42) and (2.40), this implies that the divergence occurs when

$$\frac{M}{R_s} = \frac{4}{9} , \qquad (2.48)$$

which is precisely the Buchdahl limit for the compactness of a star in GR. In turn, since we expect to be able to describe configurations below the Buchdahl limit, with $\gamma > 1$, Eq. (2.47) implies that $\mathcal{U}^{(-)}(0)$ is always negative in this regime of compactness. Moreover, $\gamma > 1$ implies that $C < 2/3\, R_s$ and $C\, r < 1$ for all values of $0 \leq r \leq R_s$. A numerical analysis also shows that the function $g(r)$ is always positive, and that $\mathcal{U}^{(-)}$ is always negative, except close to the surface of a star with a compactness near the limit (2.48), as shown in Fig. 2.4 for $R_s = 5$ and $C = 0.43$ ($M/R_s \approx 0.42$).

The behaviour of the anisotropic term $\mathcal{P}^{(-)}$ at $r = 0$ depends on the function $g(r)$ defined in (2.41). For the latter one finds

$$\lim_{r \to 0} \frac{g(r)}{r^2} = \frac{1}{6\,(\gamma - 1)} , \qquad (2.49)$$

hence $\mathcal{P}^{(-)}(0) = 0$. Figure 2.3 shows $\mathcal{P}^{(-)}$ inside the stellar distribution for different densities. It can be seen that the anisotropic stress is proportional to the density (which is proportional to C): the more compact distributions display larger

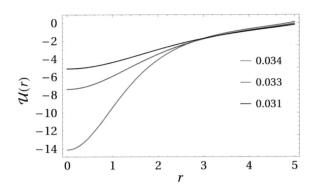

Fig. 2.4 Schwarzschild solution. Weyl function $[\mathcal{U}^{(-)} \times 10^3]$ for three different values of C with $R_\mathrm{s} = 5$

anisotropic effects. This behaviour is easily explained in terms of the source of the anisotropy, which is nothing but the geometric deformation undergone by the radial metric component, explicitly shown in the solution (1.56). When $H = 0$, the only source for the geometric deformation are the high-energy terms in the expression (1.56), which are quadratic in the density and pressure. Hence the higher the density, the larger the geometric deformation and, consequently, the anisotropy will be larger for more compact distributions. However, this picture changes near the surface, where the opposite behaviour occurs again because of the matching conditions with the exterior. One can also see that the anisotropy increases inward from the surface until it reaches a maximum value, and then it decreases until $\mathcal{P}^{(-)} = 0$ at $r = 0$, in agreement with the behaviour of the geometric deformation (1.60).

On the other hand, Fig. 2.4 shows the Weyl function $\mathcal{U}^{(-)}$ for the three distributions considered in Fig. 2.3. This function is negative and larger for more compact stellar objects except for the external layers. This means that, inside the star, high-energy terms always dominate over anisotropic terms, which are the two sources for $\mathcal{U}^{(-)}$, as can be seen from Eq. (1.37). However, there appears a layer at the surface inside which $\mathcal{U}^{(-)}$ becomes positive, whose thickness increases for more compact sources.

2.2.2   Matching Conditions

We have already seen that an interior BW solutions cannot be matched with the exterior Schwarzschild solution (1.77) without additional assumptions. We shall therefore match the above interior solution for $r < R_\mathrm{s}$ with the exterior tidally charged metric (2.21) and (2.22) for $r > R_\mathrm{s}$. The matching conditions (1.70), (1.71) and (1.75) imply

$$4\,B^2\left(1 - C\,R_\mathrm{s}^2\right) = 1 - \frac{2\,\mathcal{M}}{R_\mathrm{s}} + \frac{q}{R_\mathrm{s}^2} \tag{2.50}$$

$$\mathcal{M} = M - \frac{18\,C^2\,R_\mathrm{s}\,g(R_\mathrm{s};\,C)}{k^2\,\sigma} + \frac{2\,q}{R_\mathrm{s}} \tag{2.51}$$

and

$$
\begin{aligned}
q = & \frac{3\, R_s^4 \left(64\, \pi^2 - 1\right) \left(8 - 15\, C\, R_s^2 + 7\, C^2\, R_s^4\right)}{1024\, \sigma\, \pi^3 \left(4 - 7\, C\, R_s^2 + 3\, C^2\, R_s^4\right)} \\
& -g(R_s;\, C) \frac{768\, \pi^2 + \left(5 - 896\, \pi^2\right) C\, R_s^2 + 3 \left(64\, \pi^2 - 1\right) C^2\, R_s^4}{1536\, \sigma\, \pi^3 \left(4 - 7\, C\, R_s^2 + 3\, C^2\, R_s^4\right)} \, .
\end{aligned}
\tag{2.52}
$$

Now, using Eq. (2.40), the matching condition (2.50) can be rewritten as

$$
4\, B^2 \left(1 - C\, R_s^2\right) = \left(1 - C\, R_s^2\right) + \frac{9\, C^2\, g(R_s;\, C)}{k^2\, \sigma} \, ,
\tag{2.53}
$$

which shows that bulk effects can be completely encoded in the constant B, without modifying the parameter C. In particular, to first order in $1/\sigma$, we can write

$$
B(\sigma) \simeq \frac{1}{2} + \frac{9\, C^2\, g(R_s;\, C)}{4\, k^2\, \sigma \left(1 - C\, R_s^2\right)} \equiv B_0 + \delta B(\sigma) \, ,
\tag{2.54}
$$

where $B_0 = 1/2$ is the GR value. Since the density and pressure depend only on the constant C, but not on B, there are no bulk effects on them in this solution.

We just mentioned that we can choose to have no five-dimensional effects on the pressure or density in this solution, because the parameter C may be left unaffected. However, a different choice is also possible. Let us start from the relationship (2.34) among the parameters A, B and C (and depending on the radius R_s). If A is not modified by five-dimensional effects, one necessarily obtains a correction δC for C due to the correction δB shown in (2.54), that is

$$
\delta C = 8 \left(1 - C\, R_s^2\right) \frac{\delta B}{C\, R_s^2} = \frac{18\, C\, g(R_s;\, C)}{2\, k^2\, \sigma\, C\, R_s^2} \, .
\tag{2.55}
$$

Since $g(R_s;\, C)$ is numerically shown to be always positive, the correction $\delta C > 0$. Figure 2.5 shows a qualitative comparison of the pressure in GR and in the BW when $\delta A = 0$ for the Schwarzschild solution. The pressure behaves differently from BW solutions with uniform densities where the internal non-local Weyl functions were excluded [20], thus showing that these non-local effects can play a relevant role inside the stellar distribution.

We have seen that a consistent version of the Schwarzschild interior metric can be constructed, with both local bulk terms (high-energy corrections) and non-local bulk terms (bulk Weyl curvature contributions) by means of the MGD. We found that both effects are in general proportional to the compactness of the stellar distribution and, in particular, that the induced anisotropy $\Pi \sim \mathcal{P}^{(-)}$ is larger for more compact distributions. We also found the Weyl scalar $\mathcal{U}^{(-)}$ is always negative inside the stellar distribution, although it is usually subdominant with respect to $\mathcal{P}^{(-)}$. However this

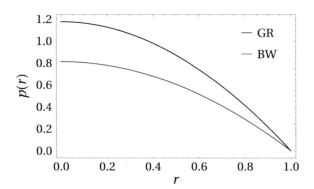

Fig. 2.5 Schwarzschild solution. Comparison between the pressure $[p \times 10^2]$ in GR and the BW with $\delta A = 0$ for a distribution with compactness $M/R_s = 0.2$

behaviour usually changes close to the surface if the compactness of the stellar distribution is near the maximum value permitted by GR.

2.3 The Durgapal-Fuloria Solution

Let us now consider an exact, and more importantly, physically acceptable solution to the system (1.27)–(1.30) starting from the spherically symmetric perfect fluid solution found by Durgapal and Fuloria in Ref. [22], namely

$$e^{\nu(-)} = A \left(1 + C \, r^2\right)^4 \tag{2.56}$$

$$k^2 \rho = \frac{8 \, C \, \left(9 + 2 \, C \, r^2 + C^2 \, r^4\right)}{7 \left(1 + C \, r^2\right)^3} \tag{2.57}$$

$$k^2 p = \frac{16 \, C \, \left(2 - 7 \, C \, r^2 - C^2 \, r^4\right)}{7 \left(1 + C \, r^2\right)^3} \, , \tag{2.58}$$

where again A and C are constants in GR, but they will be functions of the brane tension σ determined by the matching conditions in this extra-dimensional context.

Using (2.56)–(2.58) in (1.56), a regular and well defined solution for $\lambda(r)$ is obtained of the usual form (1.63), where the interior mass function $\tilde{m}$ is given by

$$\tilde{m} = m(r) - \frac{32 \, C \, r}{49 \, k^2 \, \sigma} \left[\frac{240 + 589 \, C \, r^2 - 25 \, C^2 \, r^4 - 41 \, C^3 \, r^6 - 3 \, C^4 \, r^8}{3 \left(1 + C \, r^2\right)^4 \left(1 + 3 \, C \, r^2\right)} \right.$$
$$\left. - \frac{80 \, \mathrm{arctg}(\sqrt{C} \, r)}{\left(1 + C \, r^2\right)^2 \left(1 + 3 \, C \, r^2\right) \sqrt{C} \, r} \right] . \tag{2.59}$$

The GR mass function (1.6) for this solution reads

$$m = C r^3 \frac{4 \left(3 + C r^2\right)}{7 \left(1 + C r^2\right)^2} , \qquad (2.60)$$

hence the total GR mass is given by

$$M \equiv m(R_s) = C R_s^3 \frac{4 \left(3 + C R_s^2\right)}{7 \left(1 + C R_s^2\right)^2} , \qquad (2.61)$$

with R_s the radius of the distribution.

Using (1.36) and (1.37), one obtains regular solutions for the interior Weyl functions, to wit

$$\frac{\mathcal{P}^{(-)}}{k^2 \sigma} = \frac{\left[Cr^2 \left(9Cr^2 - 10\right) - 3\right] \tilde{m}(r) + \left(5C^2r^5 + 6Cr^3 + r\right) \tilde{m}'(r) - 8C^2r^5}{6r^3 \left(Cr^2 + 1\right)^2} \qquad (2.62)$$

and

$$\begin{aligned} \frac{\mathcal{U}^{(-)}}{k^2 \sigma} &= \frac{4C \left(41C^2r^4 + 98Cr^2 + 9\right)}{21 \left(Cr^2 + 1\right)^3} \\ &\quad + \frac{16C^2 \left(3C^4r^8 + 32C^3r^6 + 62C^2r^4 + 200Cr^2 + 153\right)}{49 \, k^2 \, \sigma \left(Cr^2 + 1\right)^6} \\ &\quad - \frac{4C \left(9Cr^2 + 5\right) \tilde{m}(r)}{3r \left(Cr^2 + 1\right)^2} - \frac{\left(5Cr^2 + 1\right) \tilde{m}'(r)}{3r^2 \left(Cr^2 + 1\right)} . \end{aligned} \qquad (2.63)$$

The expressions (2.56)–(2.59) with (2.62) and (2.63) represent an *exact analytic solution* to the system (1.27)–(1.30).

As can be seen from Fig. 2.6, the function $\mathcal{U}^{(-)}$ is always negative, with a maximum negative value at the origin $r = 0$. This can again be explained by looking at the

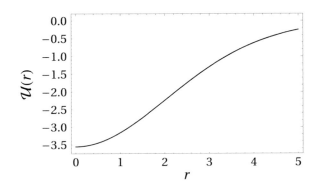

Fig. 2.6 Durgapal-Fuloria solution. Weyl function $[\mathcal{U}^{(-)} \times 10^3]$ for $R_s = 5$. It is always negative and reduces both the effective density and effective pressure

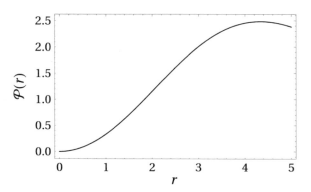

Fig. 2.7 Durgapal-Fuloria solution. Anisotropy $[\mathcal{P}^{(-)} \times 10^4]$ for $R_s = 5$

general expression for $\mathcal{U}^{(-)}$ in Eq. (1.37), which contains two kinds of "sources": the first two terms in the right hand side of Eq. (1.37) are high-energy corrections and always yield a negative contribution; the remaining terms represent an anisotropy and are always positive. Whenever the anisotropy is not large enough, the dominant high-energy terms will generate a negative $\mathcal{U}^{(-)}$, like in the case presented here. This negative scalar function in turn reduces both the effective density and the effective pressure, as can be seen from the field Eqs. (1.27)–(1.29). On the other hand, the anisotropy $\Pi \sim \mathcal{P}^{(-)}$ inside this BW star is shown in Fig. 2.7. It increases inward up to a maximum value and then decreases until $\mathcal{P}^{(-)} = 0$ at $r = 0$. This is directly connected with the correction for λ being proportional to high-energy terms as shown in Eq. (1.56). This correction is the only bulk effect on the metric when the constraint $H = 0$ is imposed, therefore it represents the only source for $\mathcal{P}^{(-)}$, as can be clearly seen in Eq. (1.36).

2.3.1   Matching Conditions

As we have already mentioned, the bulk contributions to p, ρ and $\nu_{(-)}$ are found by matching with the outer geometry, where the assumption of vanishing pressure at the surface will be dropped [15, 16]. We also know that the Schwarzschild exterior solution is incompatible with the interior solution found here (again, unless additional assumptions are considered) and a different exterior solution will therefore be considered.

Using the tidally-charged solution (2.21) and (2.22), the matching conditions in Eqs. (1.70) and (1.71) yield

$$A \left(1 + C\, R_s^2\right)^4 = 1 - \frac{2\,\mathcal{M}}{R_s} + \frac{q}{R_s^2} \qquad (2.64)$$

and

$$\mathcal{M} = M - \frac{16\,C\,R_{\rm s}}{49\,k^2\,\sigma} \left[\frac{240 + 589\,C\,R_{\rm s}^2 - 25\,C^2\,R_{\rm s}^4 - 41C^3\,R_{\rm s}^6 - 3\,C^4\,R_{\rm s}^8}{3\left(1 + C\,R_{\rm s}^2\right)^4 \left(1 + 3\,C\,R_{\rm s}^2\right)} \right.$$
$$\left. - \frac{80\arctan(\sqrt{C}\,R_{\rm s})}{\left(1 + C\,R_{\rm s}^2\right)^2 \left(1 + 3\,C\,R_{\rm s}^2\right)\sqrt{C}R_{\rm s}} \right] + \frac{q}{R_{\rm s}^2} \tag{2.65}$$

and, by using Eq. (1.75), we obtain

$$q = \frac{32\,C\,R_{\rm s}}{147\,k^4\,\sigma\,R_{\rm s}^3\left(1 + C\,R_{\rm s}^2\right)^5\left(1 + 3\,C\,R_{\rm s}^2\right)}$$
$$\times \left[\frac{21}{2}\,k^2\,\sigma\,R_{\rm s}^2\left(1 + C R_{\rm s}^2\right)^2\left(C\,R_{\rm s}^2 + 22\,C^2\,R_{\rm s}^4 + 3\,C^3\,R_{\rm s}^6 - 2\right) \right.$$
$$+ 240 + 2749\,C\,R_{\rm s}^2 + 5276\,C^2\,R_{\rm s}^4 + 266\,C^3\,R_{\rm s}^6$$
$$- 372\,C^4\,R_{\rm s}^8 - 27\,C^5\,R_{\rm s}^{10}$$
$$\left. + \frac{240\left(1 + C\,R_{\rm s}^2\right)^2\left(1 + 9\,C\,R_{\rm s}^2\right)\arctan\left(\sqrt{C}\,R_{\rm s}\right)}{\sqrt{C}\,R_{\rm s}} \right]. \tag{2.66}$$

The constant C can now be determined from Eq. (2.64) if A is kept as a free parameter. However, we should also recall that A has a specific value, say A_0, in GR which is given by taking the limit $\sigma^{-1} \to 0$ in Eq. (2.64), that is

$$A_0\left(1 + C_0\,R_{\rm s}^2\right)^4 = 1 - \frac{2\,M}{R_{\rm s}}, \tag{2.67}$$

where C_0 is the GR value of C given by

$$C_0 = \frac{\sqrt{57} - 7}{2\,R_{\rm s}^2}, \tag{2.68}$$

as follows from $p(R_{\rm s}) = 0$ in Eq. (2.2). It is important to note that the first term inside the square brackets of Eq. (2.66) apparently survives in the limit $\sigma^{-1} \to 0$. However, it exactly vanishes for $C = C_0$, as it should be in GR. On the other hand, we can assume that the effects of the bulk on C are given by the perturbative expression

$$C = C_0 + \delta C, \tag{2.69}$$

where $\delta C = \delta C(\sigma)$ is such that $\delta C(\sigma \to \infty) = 0$, as well as a similar expression for A,

$$A = A_0 + \delta A, \tag{2.70}$$

with $\delta A(\sigma \to \infty) = 0$. The matching condition (2.64) then reads

Fig. 2.8 Durgapal-Fuloria
solution. Qualitative
comparison of the pressure
$[p \times 10^2]$ in GR and in the
BW model, with $R_s = 5$

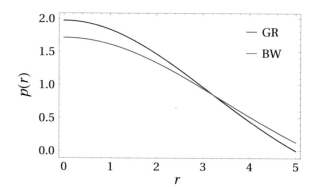

$$(A_0 + \delta A) \left[1 + (C_0 + \delta C) R_s^2\right]^4 = 1 - \frac{2\mathcal{M}}{R_s} + \frac{q}{R_s^2}. \tag{2.71}$$

The expressions (2.65) and (2.66) can now be evaluated to linear order in σ^{-1} and inserted into Eq. (2.71) leading to

$$\delta C = \frac{7\left(1 + C_0 R_s^2\right)^3 \left[\alpha(\sigma) - \left(1 + C_0 R_s^2\right)^4 \delta A\right]}{4\left(7 + 2 C_0 R_s^2 + C_0^2 R_s^4 - 2 C_0^3 R_s^6\right)}, \tag{2.72}$$

with

$$\alpha = \frac{32\, C_0}{49\, k^2\, \sigma} \left[\frac{240 + 589\, C_0\, R_s^2 - 25\, C_0^2\, R_s^4 - 41\, C_0^3\, R_s^6 - 3\, C_0^4\, R_s^8}{3\left(1 + C_0\, R_s^2\right)^4 \left(1 + 3\, C_0\, R_s^2\right)}\right.$$
$$\left. - \frac{80 \arctan(\sqrt{C_0}\, R_s)}{\left(1 + C_0\, R_s^2\right)^2 \left(1 + 3\, C_0\, R_s^2\right) \sqrt{C_0}\, R_s}\right]. \tag{2.73}$$

Hence, given the perturbation δA, it is possible to obtain δC, which determines the effects of the bulk on the pressure p and density ρ via Eqs. (2.18) and (2.19). Figure 2.8 shows the behaviour of the pressure in GR and in the BW. It can be seen that bulk effects reduce the pressure deep inside the distribution, but the situation changes for the outer layers, where the matching conditions lead to $p \neq 0$ at the surface.

2.4 Tolman IV Solution

We shall now build one more solution to the system of Eqs. (1.27)–(1.30) by generating the BW version of the Tolman IV solution [23] in exact analytical form. The reason to investigate this particular case by means of the MGD approach is that, among the known exact solutions in GR, the Tolman IV solution is one of the few with physically sensible properties. In fact, it is regular in the origin; the corresponding pressure and density are everywhere positive, satisfy the dominant energy condition and decrease monotonically from the centre outward; the mass and radius of the star are well defined the speed of sound is subluminal, etc. We will then see that these properties are naturally inherited by the BW version obtained from the MGD approach.

Let us begin by considering the Tolman IV solution for a perfect fluid in GR deformed by five-dimensional effects through the function f^*, that is the set of quantities

$$e^{\nu_{(-)}} = B^2 \left(1 + \frac{r^2}{A^2} \right) \tag{2.74}$$

$$e^{-\lambda_{(-)}} = \frac{(C^2 - r^2)(A^2 + r^2)}{C^2 (A^2 + 2r^2)} + f^*(r) \tag{2.75}$$

$$k^2 \rho = \frac{3A^4 + A^2(3C^2 + 7r^2) + 2r^2(C^2 + 3r^2)}{C^2(A^2 + 2r^2)^2} \tag{2.76}$$

$$k^2 p = \frac{C^2 - A^2 - 3r^2}{C^2(A^2 + 2r^2)} . \tag{2.77}$$

In GR, i.e. when $f^* = 0$, the parameters A, B and C take specific values which can be written in terms of the compactness M/R of the distribution. In turn, the mass M and the radius R of the distribution can be freely assigned provided they satisfy the constraint $M/R < 4/9$ [the Buchdahl limit, see further bellow Eqs. (2.88)–(2.90)]. However, in the BW scenario the matching conditions are modified and there are five-dimensional effects on these constants which must consequently be considered. Indeed, as we have already seen in the previous cases, A, B and C will in general be functions of the brane tension σ which are determined by the matching conditions, and reduce to the GR values for $1/\sigma \to 0$. More clearly, as long as the brane tension σ remains uniform and constant, A, B and C will not vary with the space-time position but will depend on M, R and σ.

Continuity of the second fundamental form in GR leads to $p(R_s) = 0$ at the surface $r = R_s$ and, from Eq. (2.77), this implies

$$C^2 = A^2 + 3R_s^2 , \tag{2.78}$$

which allows us to eliminate the parameter C from all expressions. Unlike previous cases, we will keep the condition that the pressure vanishes at the surface and determine its consequences in the BW. In particular, the minimal deformation f^* for the Tolman IV solution can be computed using Eqs. (2.74), (2.76) and (2.77) in Eq. (1.60), and explicitly reads

$$f^* = \frac{A^2 + r^2}{k^2 \, \sigma \, S(r)} \left\{ \frac{3\,r\,\sqrt{2\,A^2 + 3\,r^2}}{\left(A^2 + 2\,r^2\right)^3} \left[5\,A^8 + 7\,A^6\,r^2 + 10\,A^2\,r^6 + 12\,r^8 \right. \right.$$

$$+ 4\left(6\,A^6 + 10\,A^4\,r^2 - 3\,A^2\,r^4 - 6\,r^6\right) R_{\mathrm{s}}^2$$

$$\left. + 2\left(15\,A^4 + 35\,A^2\,r^2 + 18\,r^4\right) R_{\mathrm{s}}^4 \right]$$

$$- 18\left(A^2 + 2\,R^2\right)^2 \arctan\left(\frac{r}{\sqrt{2\,A^2 + 3\,r^2}}\right)$$

$$\left. - 4\,\sqrt{3}\left(A^2 + 3\,R^2\right)^2 \ln\left(3\,r + \sqrt{6\,A^2 + 9\,r^2}\right) \right\} \tag{2.79}$$

where

$$S = -4\,r\left(A^2 + 3\,R_{\mathrm{s}}^2\right)^2 \left(2\,A^2 + 3\,r^2\right)^{3/2}. \tag{2.80}$$

The GR mass function in Eq. (1.64) is also given by

$$m = \frac{r^3\left(2\,A^2 + 3\,R^2 + r^2\right)}{2\left(A^2 + 3\,R^2\right)\left(A^2 + 2\,r^2\right)}, \tag{2.81}$$

and

$$M \equiv m(R_{\mathrm{s}}) = \frac{R_{\mathrm{s}}^3}{A^2 + 3\,R_{\mathrm{s}}^2}. \tag{2.82}$$

Finally, the Weyl functions associated with the geometric deformation (2.79) are written as

$$\frac{\mathcal{P}^{(-)}}{k^4\,\sigma} = \frac{\left(A^4 + 2\,A^2\,r^2 + 2\,r^4\right)}{6\,r^2\left(A^2 + r^2\right)^2}\,f^*(r) \tag{2.83}$$

and

$$\frac{\mathcal{U}^{(-)}}{k^2\,\sigma} = \frac{k^2\left(A^4 + 8\,A^2\,r^2 + 5\,r^4\right)}{6\,r^2\left(A^2 + r^2\right)^2}\,f^*(r)$$

$$- \frac{9\left(2\,A^4 + 3\,A^2\,r^2 + 2\,r^4 + 3\,A^2\,R_{\mathrm{s}}^2 + 2\,r^2\,R_{\mathrm{s}}^2\right)}{4\,k^2\,\sigma\left(A^2 + 2\,r^2\right)^4}$$

$$\times \frac{2\,A^4 + A^2\,r^2 - 2\,r^4 + 5\,A^2\,R_{\mathrm{s}}^2 + 6\,r^2\,R_{\mathrm{s}}^2}{\left(A^2 + 3\,R_{\mathrm{s}}^2\right)^2}. \tag{2.84}$$

The expressions (2.74)–(2.77) along with Eqs. (2.83) and (2.84) represent an exact analytic solution to the system of Eqs. (1.27)–(1.30). Since p, ρ and $\nu_{(-)}$ are exactly the same as those for the Tolman IV solution, the condition $H = 0$ in Eq. (1.59) is satisfied. One important aspect is that the minimal geometric deformation f^* in Eq. (2.79) depends only on the parameter A when Eq. (2.78) holds, which has a well defined expression in terms of the compactness M/R in GR, as will be shown next.

2.4.1 General Relativity Limit

In order to clarify the physical consequences due to BW effects, we first review
the GR case by matching the above interior Tolman IV solution with the exterior
Schwarzschild metric, that is

$$e^{\nu_{(+)}} = e^{-\lambda_{(+)}} = 1 - \frac{2M}{r} .$$

(2.85)

At the stellar surface $r = R_{\rm s}$ we therefore have, according to Eqs. (1.70) and (1.71),

$$B^2 \left(1 + \frac{R_{\rm s}^2}{A^2}\right) = 1 - \frac{2M}{R_{\rm s}}$$

(2.86)

and

$$\frac{\left(C^2 - R_{\rm s}^2\right)\left(A^2 + R_{\rm s}^2\right)}{C^2 \left(A^2 + 2 R_{\rm s}^2\right)} = 1 - \frac{2M}{R_{\rm s}} .$$

(2.87)

Moreover, the continuity condition (1.76) with $f_{R_{\rm s}}^* = \mathcal{U}_{R_{\rm s}}^{(+)} = \mathcal{P}_{R_{\rm s}}^{(+)} = 0$ leads to
$p(R_{\rm s}) = 0$, which is equivalent to Eq. (2.78), as we have just seen above. By using
Eq. (2.78) along with Eqs. (2.86) and (2.87) the constants A, B and C can be expressed
in term of the compactness of the stellar distribution. Like before, we denote these
GR expressions as

$$\frac{A_0^2}{R_{\rm s}^2} = \frac{R_{\rm s} - 3M}{M}$$

(2.88)

$$B_0^2 = 1 - \frac{3M}{R_{\rm s}}$$

(2.89)

$$\frac{C_0^2}{R_{\rm s}^2} = \frac{R_{\rm s}}{M} .$$

(2.90)

We will see next that the BW case can be quite different even if we keep the condition (2.78) and $p(R_{\rm s}) = 0$.

2.4.2 Matching Conditions

When the deformed Tolman IV interior solution, given by Eqs. (2.74) and (2.75), is
used along with the tidally-charged solution (2.21) in the matching conditions (1.70)
and (1.71), we find

$$B^2 \left(1 + \frac{R_{\rm s}^2}{A^2}\right) = 1 - \frac{2\mathcal{M}}{R_{\rm s}} - \frac{q}{R_{\rm s}^2} ,$$

(2.91)

$$\frac{2\,\mathcal{M}}{R_\mathrm{s}} = \frac{2\,M}{R_\mathrm{s}} - f_{R_\mathrm{s}}^* - \frac{q}{R_\mathrm{s}^2} \,, \tag{2.92}$$

where the geometric deformation $f_{R_\mathrm{s}}^*$ is given by Eq. (2.79) evaluated at the star surface $r = R_\mathrm{s}$. The tidal charge q can be expressed in terms of $f_{R_\mathrm{s}}^*$ by making use of Eq. (2.22) in the second fundamental form (1.76),

$$\frac{q}{R_\mathrm{s}^4} = \left(\frac{\nu'_{R_\mathrm{s}}}{R_\mathrm{s}} + \frac{1}{R_\mathrm{s}^2}\right) f_{R_\mathrm{s}}^* \,, \tag{2.93}$$

which, for our interior solution (2.74), becomes

$$\frac{q}{R_\mathrm{s}^3} = \frac{f_{R_\mathrm{s}}^*}{R_\mathrm{s} - 2\,M} \,. \tag{2.94}$$

Replacing Eq. (2.92) into Eq. (2.91) leads to

$$B^2\left(1 + \frac{R_\mathrm{s}^2}{A^2}\right) = 1 - \frac{2\,M}{R_\mathrm{s}} + f_{R_\mathrm{s}}^* \,, \tag{2.95}$$

so that the parameters A and B cannot equal the GR expressions A_0 and B_0 in Eqs. (2.88) and (2.89), since this would lead to the trivial Schwarzschild condition $f_{R_\mathrm{s}}^* = 0$. Indeed, we will next see that they must depend on the brane tension σ.

Since A and B just satisfy the one condition (2.95), additional information must be provided in order to fix their form. First of all, we note that BW effects on A and B are precisely determined by the geometric deformation $f_{R_\mathrm{s}}^*$ which appears in the right hand side of Eq. (2.95). The latter is proportional to σ^{-1} [see Eq. (2.79)] and we can therefore write

$$A = A_0 + \delta A \tag{2.96}$$
$$B = B_0 + \delta B \,, \tag{2.97}$$

where δA and δB are functions of the brane tension σ and must vanish for $\sigma^{-1} \to 0$. On the other hand, any change in the parameters A and B must be accompanied by a change in the compactness M/R_s which becomes $\mathcal{M}/R_\mathrm{s}$ in the BW. If we keep the radius R_s as a fixed free parameter, the five-dimensional effects on A and B will result in the change of the mass $\mathcal{M} = M + \delta M$. In order to identify unique values of A and B we can for instance require that the GR relations between δA and δB are the same as in GR, that is

$$A^2 = \frac{3\,R_\mathrm{s}^2\,B^2}{(1 - B^2)} \,, \tag{2.98}$$

so that the problem at the surface is closed. Using Eqs. (2.96)–(2.97) in (2.95) and (2.98), and keeping only terms to first order in δA and δB, finally yields

Fig. 2.9 Tolman IV solution. Comparison of the pressure $[p \times 10^3]$ in GR and in the BW model for a distribution with compactness $M/R_s = 0.2$

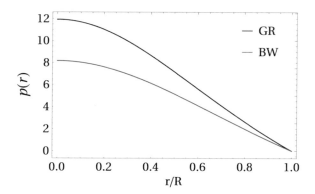

$$\delta A(\sigma) \simeq \frac{A_0^3 \, f_{R_s}^*}{4 \, R_s^2 \, B_0^4} \, , \tag{2.99}$$

which is always positive because the geometric deformation $f_{R_s}^* \geq 0$.

We can now examine the five-dimensional effects on the physical variables. For instance, we can write the pressure (2.77) as

$$p(\sigma) = \frac{3 \left(R_s^2 - r^2 \right)}{8 \, \pi \left(A^2 + 3 \, R_s^2 \right) \left(A^2 + 2 \, r^2 \right)} \, . \tag{2.100}$$

Since $A > A_0$, this BW pressure will always be smaller than the GR counterpart (see Fig. 2.9).

2.5 Brane World Stars with Solid Crust

It is commonly believed that it should be possible to detect BW modifications to a star in the weak field regime. Orbital motions or gravitational lensing could then provide information not only about the parameters of the stars like its mass, rotation and quadrupole moment, but also about BW effects. Strong field effects, like those occurring in the inner edge of an accretion disk, also depend only on the exterior geometry. Stellar astrophysical processes leading to electromagnetic radiation could also be slightly modified in the BW. Although standard model fields remain four-dimensional, gravity is changed and the equilibrium between radiation pressure and gravitational attraction, for example, is shifted.

All of the four interior solutions we have presented so far were assumed to be generated by regular fluids with positive pressure everywhere inside the star. However, the general results of Sect. 1.6 showed that BW effects induce a negative pressure at the surface of such a star if the outer geometry is given by the Schwarzschild vacuum. As a consequence, the exterior BW geometry was also modified by including a tidal charge whose effects on the motion of test particles could be detected, at least in principle, and lead to experimental bounds on the values of BW parameters. In fact, such constraints were already derived for the tidally charged metric [9], and include constraints on the tidal charge from the deflection of light [24–27], from the radius of the first relativistic Einstein ring due to strong lensing [28] and from the emission properties of the accretion disks, including the energy flux, the emission spectrum and accretion efficiency [29].

In this section, we shall instead allow for a negative pressure at the surface, which we interpreted as the presence of a solid crust in Ref. [30]. The appearance of solid materials in the BW was first advanced in Ref. [31], which presented the homogeneous counterpart of the Einstein brane [32]. Another BW star, which obeys the dark energy condition in the latest stages of the collapse, was discussed in Ref. [15]. In particular, we shall show that the stellar matter can have the most reasonable physical properties when matching conditions are obeyed between the interior geometry and the exterior Schwarzschild vacuum. The importance of this result is that it illustrates that no matter how severe the experimental constraints, from lensing or other tests, for BW stars with exteriors depending on BW parameters, the very existence of BW stars cannot be ruled out entirely, since the surrounding geometry could be exactly the same as in GR.

2.5.1 Stellar Interior

We start from the exact BW interior solution described in Sect. 2.3. By introducing the variable $x = C r^2$, that solution can be written as

$$e^{\nu(-)} = A (1 + x)^4 \, , \tag{2.101}$$

$$e^{-\lambda(-)} = 1 - \frac{8 x (3 + x)}{7 (1 + x)^2} + f^*(x) \, , \tag{2.102}$$

which is generated by the density

$$k^2 \rho = \frac{8 C (9 + 2 x + x^2)}{7 (1 + x)^3} \, , \tag{2.103}$$

and by the isotropic pressure

$$k^2 p = \frac{16 C (2 - 7 x - x^2)}{7 (1 + x)^3} \, . \tag{2.104}$$

The geometric deformation in Eq. (2.102) is now given by

$$f^* = \frac{2^2\,C}{7^2\,\sigma\,\pi}\left[\frac{240 + 589\,x - 25\,x^2 - 41\,x^3 - 3\,x^4}{3\,(1+x)^4\,(1+3x)} - \frac{80\,\arctan(\sqrt{x})}{(1+x)^2\,(1+3\,x)\sqrt{x}}\right],$$
$$(2.105)$$

and the Weyl terms read

$$\mathcal{P}^{(-)} = \frac{32\,C^2}{441x^2(1+x)^6(1+3x)^2}\left[x\left(180 + 2040\,x + 8696\,x^2 + 16553\,x^3\right.\right.$$
$$+12660\,x^4 + 146\,x^5 - 120\,x^6 + 9\,x^7)$$
$$\left.-60\,\sqrt{C}\,(1+x)^3(3 + 26\,x + 63\,x^2)\arctan(\sqrt{x})\right],\qquad(2.106)$$

and

$$\mathcal{U}^{(-)} = \frac{32\,C^2}{441\,x^2\,(1+x)^6(1+3x)^2}\left[x^2\left(795 + 4865\,x + 10044\,x^2\right.\right. \quad(2.107)$$
$$+6186\,x^3 - 373\,x^4 - 219\,x^5 - 18\,x^6)$$
$$\left.-240\,x^{3/2}\,(1+x)^3\,(5 + 9x)\arctan(\sqrt{x})\right].\qquad(2.108)$$

The constants A and C will then be determined from the continuity of the metric ensured by the matching conditions (1.70) and (1.71) with the Schwarzschild exterior geometry (1.77), which yield

$$A = \left(1 - \frac{2\mathcal{M}}{R_s}\right)\frac{1}{(1 + C\,R_s^2)^4}\qquad(2.109)$$

and

$$\frac{2\,\mathcal{M}}{R_s} = \frac{2\,M}{R_s} - \frac{2^2\,C}{7^2\,\sigma\,\pi}\,\frac{240 + 589\,C\,R_s^2 - 25\,C^2\,R_s^4 - 41\,C^3\,R_s^6 - 3\,C^4\,R_s^8}{3\,(1 + C\,R_s^2)^4\,(1 + 3\,C\,R_s^2)}$$
$$+\frac{2^2\,C}{7^2\,\sigma\,\pi}\,\frac{80\,\arctan(\sqrt{C}\,R_s)}{(1 + C\,R_s^2)^2\,(1 + 3\,C\,R_s^2)\,\sqrt{C}\,R_s}.\qquad(2.110)$$

Finally, the matching condition (1.78) for the continuity of the extrinsic curvature at the star surface reads

$$C\,R_s^2\left(2 - 7\,C\,R_s^2 - C^2\,R_s^4\right) + \frac{7}{16}\left(1 + C\,R_s^2\right)^2\left(1 + 9\,C\,R_s^2\right)f_{R_s}^* = 0,\quad(2.111)$$

which clearly shows that $C = C(\sigma)$ is now promoted to be a function of the brane tension σ due to bulk gravity effects. The Kretschmann scalar $R^{\mu\nu\sigma\rho}\,R_{\mu\nu\sigma\rho}$, Ricci square $R^{\mu\nu}\,R_{\mu\nu}$ and Weyl square $C^{\mu\nu\sigma\rho}\,C_{\mu\nu\sigma\rho}$ associated with the interior geometry given by the expressions (2.101) and (2.102) are shown in Fig. 2.10.

In order to find the extra-dimensional effects on the physical variables, i.e., the pressure (2.104) and the density (2.103), we need to fix the function $C = C(\sigma)$

Fig. 2.10 The Kretschmann
scalar, Ricci square and Weyl
square for a distribution with
$R = 13$ km

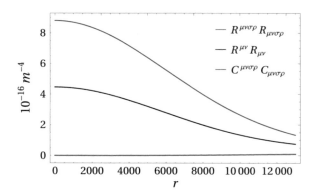

satisfying Eq. (2.111). We will do that by perturbing the seed GR solution with a
term expanded in powers of $1/\sigma$. We correspondingly write

$$C = C_0 + \delta C , \tag{2.112}$$

where the GR value of C is given by

$$C_0 = \frac{\sqrt{57} - 7}{2 R_{\mathrm{s}}^2} , \tag{2.113}$$

as follows from the standard condition $p(R_{\mathrm{s}}) = 0$ applied to the pressure (2.104). By
using Eq. (2.112) in Eq. (2.111), we then obtain the leading-order BW contribution

$$\delta C(\sigma) = \frac{7 (1 + C_0 R_{\mathrm{s}}^2)^2 (1 + 9 C_0 R_{\mathrm{s}}^2)}{16 C_0 R_{\mathrm{s}}^2 (7 + 2 C_0 R_{\mathrm{s}}^2)} \frac{f_{R_{\mathrm{s}}}^*}{R_{\mathrm{s}}^2} + \mathcal{O}(\sigma^{-2}) . \tag{2.114}$$

The pressure can then be determined by expanding $p = p(C)$ around C_0,

$$p(C_0 + \delta C) \simeq p(C_0) + \delta C \left. \frac{dp}{dC} \right|_{C=C_0} , \tag{2.115}$$

which leads to

$$k^2 p \simeq \frac{16 C_0 (2 - 7 C_0 r^2 - C_0^2 r^4)}{7 (1 + C_0 r^2)^3} + \frac{32 (1 - 9 C_0 r^2 + 2 C_0^2 r^4)}{7 (1 + C_0 r^2)^4} \delta C(\sigma) . \tag{2.116}$$

Consequently, at the star surface $r = R_{\mathrm{s}}$, the pressure becomes

$$k^2 p_{R_{\mathrm{s}}} \simeq \frac{32 (1 - 9 C_0 R_{\mathrm{s}}^2 + 2 C_0^2 R_{\mathrm{s}}^4)}{7 (1 + C_0 R_{\mathrm{s}}^2)^4} \delta C(\sigma) < 0 , \tag{2.117}$$

Fig. 2.11 Density
$[\rho(r) \times 10^{20}\,\mathrm{g/m^3}]$ for
$\delta C/C_0 = 0.03$, for a typical
compact distribution of
$R_s = 13\,\mathrm{km}$

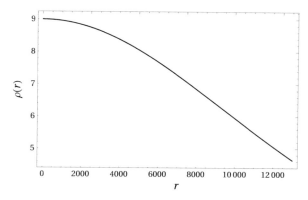

Fig. 2.12 Comparison of the
pressure $[p(r) \times 10^{19}\,\mathrm{g/m^3}]$
for a compact distribution of
$R_s = 13\,\mathrm{km}$ in GR and in the
BW model with
$\delta C/C_0 = 0.03$

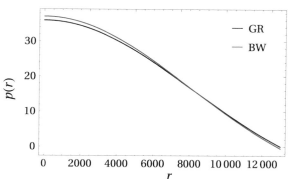

that is, negative and proportional to $1/\sigma$, according to Eqs. (2.105) and (2.114).

A typical density profile $\rho = \rho(r)$ is displayed in Fig. 2.11. In Fig. 2.12, we likewise display the pressure profile $p = p(r)$ both for the GR case and its BW generalisation. Remarkably, the pressure is negative only in a thin layer close to the surface, which acts as the positive tension typical of solid materials. The matter inside the star therefore appears as an (effectively) imperfect fluid with a solid crust.

2.5.2 Energy Conditions

In order to avoid exotic matter sources and, correspondingly, exotic space-time geometries, one can impose constraints on the energy-momentum tensor. Among these constraints, in particular, we recall:

(a) the Null Energy Condition (NEC), $T_{\mu\nu}K^\mu K^\nu \geq 0$ for any null vector K^μ. For a perfect fluid, this implies

$$\rho + p \geq 0 \; ; \tag{2.118}$$

(b) the Weak Energy Condition (WEC), $T_{\mu\nu}X^\mu X^\nu \geq 0$ for any time-like vector X^μ, which, again for a perfect fluid, yields Eq. (2.118) with further

$$\rho \geq 0 \; ; \tag{2.119}$$

(c) the Dominant Energy Condition (DEC), $T^\mu_\nu X^\nu = -Y^\mu$, where Y^μ must be a future-pointing causal vector. For a perfect fluid, this means

$$\rho > |p| \,. \tag{2.120}$$

And finally
(d) the Strong Energy Condition (SEC), $\left(T_{\mu\nu} - \frac{1}{2} T g_{\mu\nu}\right) X^\mu X^\nu \geq 0$. For a perfect fluid, this implies Eq. (2.118) and

$$\rho + 3 p \geq 0 \,. \tag{2.121}$$

Although these conditions might fail for particular classical systems which are still reasonable, they can be viewed as sensible guidelines to avoid unphysical solutions. An example is provided by classical fields obeying the WEC, which rules out wormholes, superluminal travel, and time machines. On the other hand, the SEC is violated by the cosmological inflation driven by a minimally coupled massive scalar field and by the accelerating universe [33]. Let us also recall that DEC $\Rightarrow$ WEC $\Rightarrow$ NEC and SEC $\Rightarrow$ NEC (but SEC does not imply WEC). We will now show how to implement these conditions in our case, where a GR isotropic fluid has been transformed into an anisotropic one by extra-dimensional effects.

We shall in particular consider the weakest NEC and compare the BW case directly to a perfect fluid. For this purpose, we need a null vector field which, in the interior metric (1.62), can be written as

$$K^\mu = e^{-\nu_{(-)}/2} \delta_0{}^\mu + e^{-\lambda(-)/2} \delta_1{}^\mu \,, \tag{2.122}$$

so that the NEC reads

$$T_{\mu\nu} K^\mu K^\nu = e^{\nu(-)} \tilde{\rho} K^0 K^0 + e^{\lambda(-)} \tilde{p}_r K^1 K^1 = \tilde{\rho} + \tilde{p}_r \geq 0 \,, \tag{2.123}$$

which looks like the standard condition (2.118) with $\rho \to \tilde{\rho}$ and $p_r \to \tilde{p}_r$. In the same way, the DEC leads to $\tilde{\rho} \geq \tilde{p}_r$ and $\tilde{\rho} \geq \tilde{p}_t$, which are precisely the analogues of Eq. (2.120).

Given the definitions for the effective density and pressures in Eqs. (1.31)–(1.33), these inequalities turn into new bounds for the prefect fluid density and pressure, that is

$$\rho \geq p + \frac{1}{\sigma}\left[\rho\, p + \frac{4}{k^4}(\mathcal{P} - \mathcal{U})\right] \,, \tag{2.124}$$

and

Fig. 2.13 The scalar Weyl function $\mathcal{U}(r)/\sigma$ [$\times 10^{-25}$ g/m^3] inside the stellar distribution with $R_s = 13$ km. is always negative in the interior, thus reducing the effective density and effective pressure

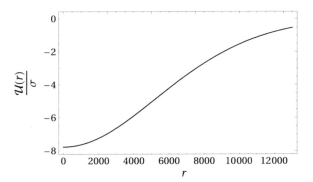

$$\rho \geq p + \frac{1}{\sigma}\left[\rho\, p - \frac{2}{k^4}\,(\mathcal{P} + 2\mathcal{U})\right],\qquad (2.125)$$

while the effective SEC becomes

$$\rho + 3\,p + \frac{1}{\sigma}\left(2\,\rho^2 + 3\,\rho\, p + \frac{12}{k^4}\,\mathcal{U}\right) \geq 0.\qquad (2.126)$$

All of the above effective conditions are satisfied, as well as the WEC, even inside the solid crust. This means that there are no negative (fluid or effective) pressures comparable in magnitude or larger than the density ρ, and therefore the BW effects on the seed GR solution are not strong enough to jeopardise the physical acceptability of the system. In Figs. 2.13 and 2.14, we display the behaviour of $\mathcal{U} = \mathcal{U}(r)$ and $\mathcal{P} = \mathcal{P}(r)$, respectively, for the same star as in Fig. 2.12. These two plots clearly show the typical energy scale of the Weyl functions and a discontinuity at $r = R_s$ in the respective quantities. We shall have more to say about this in the next section.

Fig. 2.14 The Weyl function $\mathcal{P}(r)/\sigma$ [$\times 10^{-26}$ g/m^3] inside the stellar distribution with $R_s = 13$ km

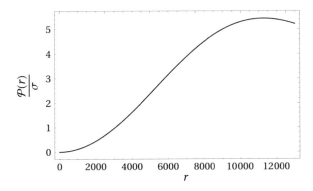

2.5.3 The Solid Crust

A test particle located on the star surface at $r = R_s$ would experience a combination
of a negative pressure $p(R_s) < 0$ and gravitational pull directed inwards, and an
extra-dimensional Weyl force pushing it outward. The negative pressure of the crust
resembles the fluid tension in a soap bubble and is analogous to the solid crust
surrounding the superfluid interior of neutron stars. Now, an important property of
this solid crust is certainly given by its thickness Δ.

In order to determine Δ, we define r_c as the value of the areal radius of the sphere
on which the pressure vanishes (see Fig. 2.12 for an example),

$$p(r_c) = 0 . \tag{2.127}$$

The (areal) thickness of the crust is thus given by

$$\Delta \equiv R_s - r_c , \tag{2.128}$$

where r_c can now be found from the pressure (2.104) with $C = C(\sigma)$ given in
Eq. (2.112). The condition (2.127) yields

$$r_c = \sqrt{\frac{\sqrt{57} - 7}{2\,C(\sigma)}} , \tag{2.129}$$

which, to leading order in $1/\sigma$, reads

$$r_c \simeq R_s \left[1 - \frac{\delta C(\sigma)}{2\,C_0} \right] , \tag{2.130}$$

so that

$$\Delta \simeq \frac{R_s\,\delta C(\sigma)}{2\,C_0} . \tag{2.131}$$

We can further rely on Eqs. (2.113), (2.114) and (2.105) in order to finally obtain

$$\Delta \sim R_s^3\,\delta C(\sigma) \sim R_s\,f_{R_s}^* \sim \frac{1}{R_s\,\sigma} . \tag{2.132}$$

Equation (2.132) shows that the solid crust is thicker the smaller the size of the
star, showing that this effect on the outer layer due to the existence of an extra-
dimension should be particularly important for very compact distributions. In fact,
for stars of solar size $R_s \gg \sigma^{-1/2}$ and the crust is much thinner than the fundamental
length $\sigma^{-1/2}$, which suggests that the crust is of little physical relevance for most
astrophysical objects, and could very possibly be eliminated by modifications of the
fundamental gravitational theory above the BW energy scale σ.

> **A BW compact source may have a Schwarzschild exterior.**

One final comment is that the stars with solid crust discussed in this section are embedded in vacuum and do not radiate (see Fig. 2.15). In order to include radiation, one could match the star interior with a Vaidya exterior containing radiation in the geometrical optics limit (or null dust). However, this would require an extension of the MGD approach to time-dependent systems, which is most likely impossible. The emission of thermal radiation can still be approximately accommodated within our approach, similarly as our Sun (the exterior of which is also well approximated by a vacuum Schwarzschild or Kerr solution) can be well described like a black body emitting radiation at approximately 5800 K. One can then argue that a black body radiation outside our BW stars with solid crust should not affect the geometry significantly, precisely like this radiation is negligible for the Sun in four-dimensional GR. Finally, the crust should be transparent to this thermal radiation, otherwise it would accumulate energy and become quickly unstable.

2.6 Classical Tests of General Relativity

Classical tests in the Solar system can probe BW signatures. The perihelion precession of Mercury, the deflection of light by the Sun and the radar echo delay observations are well-known tests for the Schwarzschild solution of GR and, in BW models, for the tidally charged and the Casadio-Fabbri-Mazzacurati metrics as well. BW effects in spherically symmetric space-times were comprehensively studied, e.g., in Ref. [34]. We will here summarise the results from Ref. [35] in which Solar system tests were employed to bound the deformation parameter β.

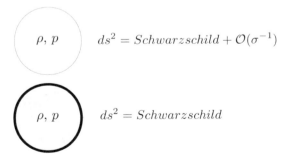

Fig. 2.15 Schematic picture of a BW star with and without a solid thin outer crust: the interior is characterised by the density ρ, pressure p and radius R_s; the exterior geometry by the ADM mass $\mathcal{M}$ and corrections proportional to σ^{-1} when there is no solid crust, and by the Schwarzschild geometry when there is a solid crust which thickness $\Delta \sim 1/R\,\sigma$

We have seen that the brane self-gravity, encoded in the brane tension σ, is one of the fundamental parameters appearing in all BW models, with $\sigma^{-1/2}$ playing the role of the (five-dimensional) fundamental gravitational length scale. In this chapter, we shall explicitly study the observational effects determined by the parameter $\beta \simeq (\sigma^{-1/2}/R_s)^2$, which describes a candidate for the modified four-dimensional geometry surrounding a star of radius R_s in the BW. By construction, the MGD approach ensures that any such BW solution smoothly reduces to the GR Schwarzschild metric in the limit $\sigma^{-1} \to 0$, thus allowing us to analyse deviations from the GR predictions for small values of the deforming parameter β.

For the study of geodesics, one only needs the deformation of the exterior spherically symmetric Schwarzschild metric (1.77) for $r > R$, where $p = \rho = 0$ but a Weyl fluid is allowed. The minimally deformed outer radial metric component will read

$$e^{\nu^+} = 1 - \frac{2M}{r} , \qquad (2.133)$$

$$e^{-\lambda^+} = \left(1 - \frac{2M}{r}\right)\left[1 + \frac{\beta(\sigma, R_s, M)}{1 - \frac{3M}{2r}}\frac{b}{r}\right] , \qquad (2.134)$$

matching the vacuum solution found in Ref. [20] when $\beta b = K/\sigma$, with $K > 0$. In the above Eq. (2.134) the (dimensionless) deformation parameter $\beta = \beta(\sigma, R_s, M)$ is the usual one that appears in the general expression (1.56), whereas

$$b \equiv R_s \left(\frac{2 R_s - 3 M}{R_s - 2 M}\right) \qquad (2.135)$$

is a length scale determined by the star radius R_s and the star compactness M/R_s. The corresponding Weyl fluid is described by [36]

$$\frac{\mathcal{P}^+}{\sigma} = \frac{\beta b \left(1 - \frac{4M}{3r}\right)}{9 r^3 \left(1 - \frac{3M}{2r}\right)^2} , \qquad (2.136)$$

and

$$\frac{\mathcal{U}^+}{\sigma} = -\frac{\beta b M}{12 r^4 \left(1 - \frac{3M}{2r}\right)^2} . \qquad (2.137)$$

The exterior geometry given by Eqs. (2.133) and (2.134) may seem to have two horizons, namely $r_h = 2M$ and $r_2 = 3M/2 - \beta b$. However, since β must be proportional to σ^{-1} in order to recover GR, the condition $r_2 < r_h$ must hold, and the outer horizon radius is given by $r_h = 2M$. The specific value $\beta = -M/2$ would produce a single horizon $r_h = r_2 = 2M$, but the limit $\sigma^{-1} \to 0$ does not reproduce the Schwarzschild solution, as seen from the condition $M_0 = M|_{\sigma^{-1} \to 0}$. On the other hand, we recall that $f_R^+ < 0$ implies that the deformed horizon radius $r_h = 2M$ is smaller than the Schwarzschild radius $r_H = 2M_0$.

We could finally obtain the parameter $\beta = \beta(\sigma, R, M)$ from the matching with an explicit internal solution. For example, the stellar interior analysed in Sect. 2.3 yields

$$\beta(\sigma, R_s) = f_{R_s}^+ = -\frac{C_0}{R_s^2\,\sigma} \,. \tag{2.138}$$

where $C_0 \simeq 1.35$ is a (dimensionless) constant. The exterior deformation to leading order in σ^{-1} finally reads

$$f^+ = -\frac{C_0\,b_0}{R_s^2\,\sigma\,r}\left(\frac{r - 2\,M_0}{r - 3\,M_0}\right) + \mathcal{O}(\sigma^{-2}) \,, \tag{2.139}$$

where $b_0 = b(M_0)$ is given by the length b at $M = M_0$. The deformation $f^+(r > R_s)$ is therefore a monotonically increasing function of the star compactness M_0/R_s. Since extra-dimensional effects are the strongest at the surface $r = R_s$ and become more important for smaller stellar distributions, the more compact the star the larger β, and thus the MGD of the GR solution.

In the following of this section we employ International System units and show explicitly the speed of light $c = 2.998 \times 10^8$ m/s and the Newton constant $G_N = 6.67 \times 10^{-11}$ m^3 kg^{-1} s^{-2}.

2.6.1 Perihelion Precession

For a test particle moving in the spherically symmetric metric (1.4) there exist two constants of motion, E and L, respectively yielding energy and angular momentum conservation. By the usual change of variable $r = 1/u$ and defining

$$g(u) = 1 - e^{-\lambda} \,, \tag{2.140}$$

the relevant equation of motion reads [34]

$$\frac{d^2u}{d\phi^2} + u = \frac{1}{2}\frac{d}{du}\left[\frac{E^2\,e^{-\lambda-\nu}}{c^2\,L^2} - \frac{e^{-\lambda}}{L^2} + g(u)\,u^2\right] \equiv k(u) \,. \tag{2.141}$$

By denoting $\gamma(u) = \left(1 - (dk/du)\,|_{u_0}\right)^{1/2}$, a circular orbit $u = u_0$ is determined by the root of the equation $u_0 = k(u_0)$, and a deviation with respect to it is provided by

$$\delta = \delta_0 \cos\left(\gamma(u)\phi + \alpha\right) \,, \tag{2.142}$$

with δ_0 and α constants. The variation of the orbital angle with respect to successive perihelia is given by

$$\phi = \frac{2\pi}{\gamma(u)} = \frac{2\pi}{1 - \Phi} \,, \tag{2.143}$$

where the perihelion advance is $\Phi \simeq \frac{1}{2} \left(\frac{dk}{du} \right)_{u=u_0}$, for small values of $(dk/du)_{u=u_0}$. For a complete rotation, the perihelion advance is $\delta\phi \leq 2\pi\Phi$.

We now consider the perihelion precession of a planet in the MGD geometry described by Eqs. (2.133) and (2.134). Since the angular momentum L is related to the semiaxis a and eccentricity e of the orbit by [34]

$$L = \frac{2\pi a^2 \sqrt{1-e^2}}{cT} , \tag{2.144}$$

where T denotes the period of motion, the perihelion advance reads

$$\delta\phi = \delta\phi_{GR} - f(\beta) , \tag{2.145}$$

where

$$\delta\phi_{\text{GR}} = \frac{6\pi G_{\text{N}} M}{c^2 a \left(1 - e^2\right)} , \tag{2.146}$$

is the well-known Schwarzschild precession formula and $f(\beta) \simeq 673.94 \, \beta$. The value for Mercury is then obtained by setting $M = M_\odot = 1.989 \times 10^{30}$ kg, $a = 57.91 \times 10^9$ m, $R_\odot = 6.955 \times 10^8$ m, and $e = 0.205615$.

If we ascribe the observed difference [37]

$$\delta\phi - \delta\phi_{\text{GR}} = 0.13 \pm 0.21 \text{ arcsec/century} \tag{2.147}$$

to BW effects, observational data [34, 37] yield the bound $f(\beta) \leq (1.89 \pm 2.33) \times 10^{-8}$, which constrains the deformation parameter to be

$$\beta \leq (2.80 \pm 3.45) \times 10^{-11} . \tag{2.148}$$

2.6.2 Light Deflection

A similar procedure describes photons on null geodesics, with the equation of motion that can be written as

$$\left(\frac{du}{d\phi} \right)^2 + u^2 = \frac{E^2 \, e^{-\nu-\lambda}}{c^2 L^2} + g(u) \, u^2 \equiv p(u) . \tag{2.149}$$

To leading order, the solution is given by

$$u = \frac{\cos\phi}{R_0}, \tag{2.150}$$

where R_0 is the distance of closest approach to the mass M. It can be iteratively employed in the above equation, yielding

$$\frac{d^2u}{d\phi^2} + u = \frac{1}{2}\frac{d}{du}\left[p\left(\frac{\cos\phi}{R_0}\right)\right]. \tag{2.151}$$

For the geometry with metric components (2.133) and (2.134), Eq. (2.140) leads to $g = \left(2\,G_{\rm N}\,M/c^2\right)u$, and for $\left(\frac{G_{\rm N}\,M}{c^2\,R_0}\right)^2 \ll 1$, $\frac{M}{L} \ll 1$ and $\frac{E^2}{c^2} - 1 \ll 1$, the total deflection of light is given by

$$\delta\phi \simeq \delta\phi_{\rm GR} + \beta\,b_0\left(\frac{E^2\,R_0}{c^2\,L^2} + \frac{18\pi c^2\,R_0}{G_{\rm N}\,M}\right), \tag{2.152}$$

where

$$\delta\phi_{\rm GR} = \frac{4\,G_{\rm N}\,M}{c^2\,R_0}. \tag{2.153}$$

Experimental uncertainties therefore imply the bound

$$\beta \le (1.07 \pm 4.28) \times 10^{-10}. \tag{2.154}$$

2.6.3 Radar Echo Delay

Another classical test of GR measures the time for radar signals to travel to, for instance, a planet [37]. The time for light to travel between two planets, respectively at a distance ℓ_1 and ℓ_2 from the Sun, is simply given by

$$T_0 = \int_{-\ell_1}^{\ell_2} \frac{dx}{c}. \tag{2.155}$$

On the other hand, if light travels in the vicinity of the Sun, or any other central star of radius $R_{\rm s}$, the time lapse $\delta T = T - T_0$ is given by [34]

$$\delta T = \int_{-\ell_1}^{\ell_2}\left\{e^{\left[\lambda\left(\sqrt{x^2+R_{\rm s}^2}\right)-\nu\left(\sqrt{x^2+R_{\rm s}^2}\right)\right]/2} - 1\right\}\frac{dx}{c}$$
$$\simeq \delta T_{\rm GR} + \frac{\beta\,n_0}{c^3\,R_{\rm s}}\left[\ln\left(\frac{4\,\ell_1\,\ell_2}{R_{\rm s}^2}\right) - \frac{5\,\pi\,G_{\rm N}M/R_{\rm s}}{2}\right], \tag{2.156}$$

where $r = \sqrt{x^2 + R_{\rm s}^2}$ and we employed a first order approximation based on Eqs. (2.133) and (2.134). Since

$$\delta T_{GR} = \frac{2\,G_N\,M}{c^3}\,\ln\left(\frac{4\,\ell_1\,\ell_2}{R_s^2}\right) \tag{2.157}$$

is the standard GR expression, the second term in Eq. (2.156) can be used to impose
a constraint on BW models. Recent measurements of the frequency shift of radio
photons both to and from the Cassini spacecraft, as they passed near the Sun, have
refined the observational constraints on the radio echo delay. For the time delay of the
signals emitted on Earth towards the Sun, one obtains $\Delta t_{radar} = \Delta t_{radar}^{GR}\,(1 + \Delta_{radar})$,
with $\Delta_{radar} \simeq (1.1 \pm 1.2) \times 10^{-5}$ [38]. In the BW geometry (2.133) and (2.134),
measurements of the frequency shift of radio photons [34, 38] yield the constraint

$$\beta \leq \frac{5\,\pi\,G_N^2\,M^2\,\Delta_{radar}}{2\,\ell_0\,R_s\,\ln\left(\frac{4\ell_1\ell_2}{R_s^2}\right)} \simeq (3.96 \pm 4.30) \times 10^{-5}\,. \tag{2.158}$$

This provides a bound on the MGD parameter β, which is the weakest one among
those in our analysis.

2.6.4 Further Remarks

BW models can be confronted with astronomical and astrophysical observations at
the Solar system scale. In this chapter we have in particular considered the BW
exterior solution (2.133) and (2.134) obtained by means of the MGD procedure,
and compared its predictions with standard GR results. This exterior geometry contains a parameter β and we were able to constrain it from the presently available
observational data in the Solar system. We found the strongest constraint is given by
measurements of the perihelion precession according to Eq. (2.148).

Let us recall that limits for the brane tension in the tidally-charged and Casadio-
Fabbri-Mazzacurati BW solutions were determined from the classical tests of GR
in Ref. [34]. Since bounds on the parameter β imply lower bounds for the brane
tension from Eq. (2.138), we can conclude that the constraint (2.148) complies with
the ones provided by such solutions of the effective four-dimensional Einstein equations (1.27)–(1.29). In fact, the brane tension in the MGD framework is bounded
according to

$$\sigma \geq \frac{9\,M_\odot\,c^2}{\pi\,R_\odot^3\,\beta}\,\frac{\left(1 - \frac{2\,G_N\,M_\odot}{c^2\,R_\odot}\right)^2}{\left(1 - \frac{3\,G_N\,M_\odot}{2\,c^2\,R_\odot}\right)}\,, \tag{2.159}$$

which implies that $\sigma \geq 5.19 \times 10^6$ MeV^4, when the bound (2.148) is taken into
account (we omit errors here since we are solely interested in orders of magnitude).
This bound is still much stronger than the constraint from cosmological nucleosyn-

thesis, albeit much weaker than the lower bound obtained from measurements of the Newton law at short scales. We can therefore conclude that the MGD geometry (2.133) and (2.134) is acceptable within the present measurements of high-energy corrections.

References

1. D. Kramer, H. Stephani, E. Herlt, M. MacCallum, *Exact Solutions of Einstein's Field Equations*(Cambridge University Press, Cambridge, 1980)
2. M.S.R. Delgaty, K. Lake, Physical acceptability of isolated, static, spherically symmetric, perfect fluid solutions of Einstein's equations. Comput. Phys. Commun. **115** 395–415 (1998)
3. A. Viznyuk, Y. Shtanov, Spherically symmetric problem on the brane and galactic rotation curves. Phys. Rev. D **76**, 064009 (2007)
4. J. Campbell, *A Course of Differential Geometry* (Clarendon, Oxford, 1926); L. Magaard, Ph.D. thesis, University of Kiel, 1963
5. S.S. Seahra, P.S. Wesson, Application of the Campbell-Magaard theorem to higher-dimensional physics. Class. Quant. Grav. **20**, 1321 (2003)
6. M.D. Maia, Hypersurfaces of Five Dimensional Space-times. arXiv:gr-qc/9512002v2
7. F. Dahia, C. Romero, On the embedding of branes in five-dimensional spaces. Class. Quant. Grav. **21**, 927–934 (2004)
8. J. Ovalle, Searching exact solutions for compact stars in braneworld: a conjecture. Mod. Phys. Lett. A **23**, 3247 (2008)
9. N. Dadhich, R. Maartens, P. Papadopoulos, V. Rezania, Black holes on the brane. Phys. Lett. B **487**, 1–6 (2000)
10. R. Casadio, A. Fabbri, L. Mazzacurati, New black holes in the brane world? Phys. Rev. D **65**, 084040 (2002)
11. P. Figueras, T. Wiseman, Gravity and large black holes in Randall-Sundrum II braneworlds. Phys. Rev. Lett. **107**, 081101 (2011)
12. D.-C. Dai, D. Stojkovic, Analytic solution for a static black hole in RSII model. Phys. Lett. B **704**, 354 (2011)
13. S. Abdolrahimi, C. Cattoen, D.N. Page, S. Yaghoobpour-Tari, Large Randall-Sundrum II black holes. Phys. Lett. B **720**, 405 (2013)
14. H. Heintzmann, New exact static solutions of Einsteins field equations. Z. Phys. **228**, 489 (1969)
15. Á. László, Gergely: black holes and dark energy from gravitational collapse on the brane. JCAP **0702**, 027 (2007)
16. N. Deruelle, Stars on branes: the view from the brane. arXiv:gr-qc/0111065v1
17. R. Casadio, J. Ovalle, Brane-world stars and (microscopic) black holes. Phys. Lett. B **715**, 251 (2012)
18. R. Casadio, J. Ovalle, Brane-world stars from minimal geometric deformation, and black holes. Gen. Relat. Grav. **46**, 1669 (2014)
19. S. Weinberg, *Gravitation and Cosmology* (Wiley, New York, 1972), p. 330
20. C. Germani, R. Maartens, Stars in the braneworld. Phys. Rev. D **64**, 124010 (2001)
21. J. Ovalle, The Schwarzschild's braneworld solution. Mod. Phys. Lett. A **25**, 3323 (2010)
22. M.C. Durgapal, R.S. Fuloria, Analytic relativistic model for a superdense star. Gen. Rel. Grav. **17**, 671 (1985)
23. R.C. Tolman, Static solutions of Einstein's field equations for spheres of fluid. Phys. Rev. **55**, 364–373 (1939)
24. C.G. Boehmer, T. Harko, F.S.N. Lobo, Solar system tests of brane world models. Class. Quant. Grav. **25**, 045015 (2008)

25. A. Kotrlova, Z. Stuchlik, G. Török, Quasiperiodic oscillations in a strong gravitational field around neutron stars testing braneworld models. Class. Quant. Grav. **25** 225016 (2008)
26. L.Á. Gergely, Z. Keresztes, M. Dwornik, Second-order light deflection by tidal charged black holes on the brane. Class. Quant. Grav. **26**, 145002 (2009)
27. Z. Horváth, L.Á. Gergely, D. Hobill, Image formation in weak gravitational lensing by tidal charged black holes. Class. Quant. Grav. **27**, 235006 (2010)
28. Z. Horváth, L.Á. Gergely, Astron. Notes (Astronomische Nachrichten) **334**, 1047 (2013)
29. C.S.J. Pun, Z. Kovacs, T. Harko, Thin accretion disks in $f(R)$ modified gravity models. Phys. Rev. D **78**, 024043 (2008)
30. J. Ovalle, L.A. Gergely, R. Casadio, Brane-world stars with solid crust and vacuum exterior. Class. Quant. Grav. **32**, 045015 (2015)
31. L.Á. Gergely, A Homogeneous brane world universe. Class. Quant. Grav. **21**, 935 (2004)
32. L.Á. Gergely, R. Maartens, Brane-world generalizations of the Einstein static universe. Class. Quant. Grav. **19**, 213 (2002)
33. M. Visser, General relativistic energy conditions: The Hubble expansion in the epoch of galaxy formation. Phys. Rev. D **56**, 7578 (1997)
34. C.G. Boehmer, G. De Risi, T. Harko, F.S.N. Lobo, Classical tests of GR in brane world models. Class. Quant. Grav. **27**, 185013 (2010)
35. R. Casadio, J. Ovalle, R. da Rocha, Classical tests of GR: brane-world sun from minimal geometric deformation. Europhys. Lett. **110**, 40003 (2015)
36. J. Ovalle, F. Linares, A. Pasqua, A. Sotomayor, The role of exterior Weyl fluids on compact stellar structures in Randall-Sundrum gravity. Class. Quant. Grav. **30**, 175019 (2013)
37. I.I. Shapiro et al., Fourth test of GR—new radar result. Phys. Rev. Lett. **26**, 1132 (1971)
38. R.D. Reasenberg, I.I. Shapiro, P.E. MacNeil, R.B. Goldstein, J.C. Breidenthal, J.P. Brenkle, D.L. Cain, T.M. Kaufman et al. Viking relativity experiment: verification of signal retardation by solar gravity. Astrophys. J. **234** L219–L221 (1979)

Chapter 3
Microscopic Black Holes

There are general arguments based on quantum physics supporting the idea that classical black holes of the kind predicted by GR can only exist with a mass significantly larger than the Planck scale $M_P = \sqrt{\hbar c / G_N}$ (where c is the speed of light, G_N is again Newton's constant and $\hbar$ the reduced Planck constant). Since $M_P \simeq 10^{-5}$ g $\simeq 10^{16}$ TeV, this means that producing black holes is utterly impossible in any foreseeable experiment on Earth. However, gravity theories have been proposed in the last decades according to which the fundamental scale of gravity, henceforth denoted by M_G, might be much lower than M_P. One of these theories is precisely the RS scenario [1, 2] reviewed in Sect. 1.2, which was in fact introduced to address the so called hierarchy problem of the Standard Model of particle physics and allows for M_G to take a value as low as the electroweak scale at 1 TeV. This model would therefore allow for the production of TeV-scale black holes at the Large Hadron Collider (LHC) [3], where however no evidence of such objects has been detected.

From the theoretical point of view, the topic is very much complicated by the non-linearity of the theory, which makes solving the five-dimensional Einstein field equations with a (point-like) source (the black hole) on top of the brane vacuum stress tensor a formidable task [4–8]. In fact, fully analytical metrics describing black holes in the complete five-dimensional space-time of the RS model are yet to be found. Moreover, the collapse of a homogeneous star in the BW was shown to be only compatible with a non-static exterior in Ref. [9], and an explicitly radiative exterior was later found [10]. This suggested that four-dimensional black holes emitting Hawking radiation [11] as a quantum effect would be equivalent to classically unstable BW black holes [12, 13]. Later on, however, the existence of large static black holes was established in Ref. [6] (see also Refs. [7, 8] for more explicit examples), and it is probably fair to say that the debate is still open as to what we should consider a realistic physical description of black holes in models with extra spatial dimensions.

© The Author(s), under exclusive license to Springer Nature Switzerland AG 2020
J. Ovalle and R. Casadio, *Beyond Einstein Gravity*, SpringerBriefs in Physics,
https://doi.org/10.1007/978-3-030-39493-6_3

In any case, it seems reasonable that a static metric (both in GR and in the BW) is an approximate description of black holes for sufficiently short times. Based on this observation, we shall consider an explicit description for a static star of radius R_s in the BW and then take the limit in which this radius (formally) goes to zero (or at least approaches values much shorter than the gravitational radius of the source). This rather formal approach will not only allow us to recover the full black hole geometry of the tidally charged metric [14] we have employed in the previous Chapter, but it will also provide a relation which uniquely determines the outer tidal charge q and ADM mass $\mathcal{M}$ of the black hole from the proper mass M of the source and the brane density σ. We should however stress that uniqueness is only achieved by making arbitrary choices and the main one we will employ is to require a linear relation between the tidal charge q and the ADM mass $\mathcal{M}$. Finally, this relation will allow us to estimate the minimum mass M_c for tidally charged black holes in terms of M_G. This result is particularly important from the phenomenological point of view, as it sets the lower limit for the energy necessary to produce microscopic black holes in high energy processes, both in our laboratories and in astrophysical systems.

In this Chapter, we shall explicitly display the Newton constant $G_N = \ell_P/M_P$ and denote by $\ell_G \gg \ell_P$ and $M_G \ll M_P$ the fundamental five-dimensional length and mass (we recall that $\hbar = M_P \ell_P = M_G \ell_G$).[1]

3.1   Regular Brane-World Stars and Black Holes

We already anticipated in Sect. 1.2 that one of the main technical issues indering the description of black holes in the BW scenario is that the four-dimensional brane on which matter is confined is usually modelled as a thin surface generated by a singular vacuum energy density proportional to $\delta(y)$ along the fifth dimension charted by the coordinate y. Therefore, if we insisted in describing a (four-dimensionally spherical) black hole as generated by a point-like source proportional to $\delta^{(3)}(\mathbf{x}) = \delta(r)/4\pi r^2$, we would end up dealing with a five-dimensional energy-momentum tensor containing the product $\delta(y)\delta(r)$. Such a contribution is mathematically ill-defined, and one should therefore regularise at least one of the two Dirac distributions by considering either a finite-thickness brane or an extended matter source for the black hole (of course, the most realistic approach would be to consider both the brane and the matter source of finite thickness).

We shall here follow the latter option and start by considering a BW star of radius R_s of the kind analysed in the previous chapter. This will let us employ the effective four-dimensional Eq. (1.14) and the MGD approach to build up an explicit solution for the interior and exterior of the star. Starting from that regular solution, we will then take the limit $R_s \to 0$ in order to obtain an effective description of BW black holes.

---

[1]Note the brane density σ has dimensions of an inverse squared length, namely $\sigma \simeq \ell_G^{-2}$.

3.2 Exterior Geometry

Like in the previous Chapter, we will describe the exterior of the source with (initially finite) radius R_s by means of the tidally charged metric [14]

$$ds^2 = e^{\nu_{(+)}} \, dt^2 - e^{\lambda_{(+)}} \, dr^2 + r^2 \left(d\theta^2 + \sin^2 \theta \, d\phi^2 \right) \, , \tag{3.1}$$

with $\lambda_{(+)} = -\nu_{(+)}$ and

$$e^{\nu_{(+)}} = 1 - \frac{2 \, \ell_P \, \mathcal{M}}{M_P \, r} - \frac{q}{r^2} \, . \tag{3.2}$$

The tidal charge q and ADM mass $\mathcal{M}$ can be treated as independent quantities when solving the effective BW vacuum Eq. (1.23). However, we have seen in the previous Chapter that matching this metric with the interior metric of a compact star will in general determine q and $\mathcal{M}$ in terms of the physical parameters characterising the interior. This (implicitly) means that q and $\mathcal{M}$ are now related. Since this relationship becomes crucial for the way we describe black holes, we first need to clarify this point further.

First of all, if no source is present and $\mathcal{M} = 0$, or in the GR limit $\sigma^{-1} \to 0$, the tidal charge q should vanish. We can therefore assume that $q = q(\mathcal{M}, \sigma)$. We then consider the junction conditions (1.70), (1.71) and (1.76) at the star surface $r = R_s$ between the tidal metric (3.1) and a general interior solution obtained from the MGD approach of the form (1.62),

$$ds_{(-)}^2 = e^{\nu_{(-)}} \, dt^2 - e^{\lambda_{(-)}} \, dr^2 - r^2 \left(d\theta^2 + \sin^2 \theta d\phi^2 \right) \, . \tag{3.3}$$

In particular, continuity of the metric (1.70) implies that $\nu_{(-)}(R_s) = \nu_{(+)}(R_s)$ and (1.71) reads

$$\frac{2 \, \mathcal{M}}{R_s} = \frac{2 \, M}{R_s} - \frac{M_P}{\ell_P} \left(f_{R_s}^* + \frac{q}{R_s^2} \right) \, , \tag{3.4}$$

where we recall that M is the total GR mass of the source and f^* the minimal deformation of the radial metric function. Further, the continuity of the extrinsic curvature (1.76) explicitly reads

$$\frac{\ell_P}{M_P} \, p_{R_s} + \left(\nu'_{R_s} + \frac{1}{R_s} \right) \frac{f_{R_s}^*}{8 \, \pi \, R_s} = \frac{2 \, \mathcal{U}_{R_s}^+}{k^4 \, \sigma} + \frac{4 \, \mathcal{P}_{R_s}^+}{k^4 \, \sigma} \, , \tag{3.5}$$

where $\nu'_{R_s} = \nu'_{(-)}(R_s)$. We can then use Eq. (3.4) and

$$\mathcal{U}_{R_s}^+ = -\frac{\mathcal{P}_{R_s}^+}{2} = \frac{4 \, \pi \, q \, \sigma}{3 \, R_s^4} \, , \tag{3.6}$$

to finally express the tidal charge as

$$\frac{M_P}{\ell_P} q = \left(\frac{R_s \nu'_{R_s} + 1}{R_s \nu'_{R_s} + 2}\right)\left(\frac{2M}{R_s} - \frac{2\mathcal{M}}{R_s}\right) R_s^2 + \frac{8\pi p_{R_s} R_s^4}{2 + R_s \nu'_{R_s}} . \qquad (3.7)$$

Since q only depends on the interior structure through ν'_{R_s}, we can determine a suitable ν'_{R_s} which lets us obtain $q = q(\mathcal{M}, \sigma)$.

For instance, by taking $p_{R_s} = 0$ and imposing the boundary constraint

$$R_s \nu'_{R_s} = -\frac{(M - \mathcal{M}) - \frac{2\mathcal{M} K M_P}{\sigma R_s^2 \ell_P}}{(M - \mathcal{M}) - \frac{\mathcal{M} K M_P}{\sigma R_s^2 \ell_P}} , \qquad (3.8)$$

where K is a (dimensionful) constant we can fix later, we obtain the linear relation [15]

$$q = \frac{2K}{\sigma R_s} \mathcal{M} , \qquad (3.9)$$

which is what we mean for a *linearly charged* black hole. Also note that the interior geometry is regular at the surface if $R_s \nu'_{R_s} > 0$, which requires

$$1 < \frac{\sigma R_s^2 \ell_P}{K M_P}\left(\frac{M - \mathcal{M}}{\mathcal{M}}\right) < 2 , \qquad (3.10)$$

as follows from Eq. (3.8). Indeed, the above linear expression for q satisfies all of our requirements, namely:

(a) it vanishes for $\mathcal{M} \to 0$ and for $\sigma^{-1} \to 0$, and
(b) it vanishes for very small star density, that is for $R_s \to \infty$ at fixed $\mathcal{M}$ and σ.

On the other hand, the fact that q diverges for $R_s \to 0$, at fixed $\mathcal{M}$ and σ, is expected from previous considerations about point-like singularities in the BW.

The exterior geometry is now fully determined by the metric function

$$e^{\nu(+)} = 1 - \frac{2\ell_P \mathcal{M}}{M_P r}\left(1 + \frac{M_P K}{\ell_P \sigma R_s r}\right) , \qquad (3.11)$$

which can also be obtained without imposing any *ad hoc* boundary constraints like the one in Eq. (3.8). We can in fact employ the result that the pressure does not need to vanish at the surface of a BW star and, from Eq. (3.7), we can find a value for p_{R_s} leading to Eq. (3.11), namely

$$4\pi R^3 p_{R_s} = \frac{M_P \mathcal{M} K}{\ell_P \sigma R_s^2}\left(2 + R_s \nu'_{R_s}\right) - (M - \mathcal{M})\left(1 + R_s \nu'_{R_s}\right) . \qquad (3.12)$$

Note that $p_{R_s} \to 0$ for $R_s \to \infty$ (at fixed M, $\mathcal{M}$ and σ), as well as for $\sigma^{-1} \to 0$ (at fixed R_s), if $\mathcal{M} \to M$ in the same limit [in fact, see Eq. (3.19) below].

3.3 Interior Geometry

The radius R_s is a free parameter so far, but it can be fixed for any specific star interiors. For instance, let us consider the exact interior BW solution described in Sect. 2.3. The temporal metric component reads

$$e^{\nu(-)} = A \left(1 + C r^2\right)^4 , \tag{3.13}$$

where A and C are constants. The associated matter density is given by

$$\rho = C_\rho \left(\frac{M_P}{\ell_P}\right) \frac{C \left(9 + 2 C r^2 + C^2 r^4\right)}{7 \pi \left(1 + C r^2\right)^3} , \tag{3.14}$$

where $C_\rho = C_\rho(K)$ is also a constant to be determined for consistency. Imposing vanishing surface pressure,

$$p_{R_s} = \left(\frac{M_P}{\ell_P}\right) \frac{2 C \left(2 - 7 C R_s^2 - C^2 R_s^4\right)}{7 \pi \left(1 + C R_s^2\right)^3} = 0 , \tag{3.15}$$

leads to

$$C = \frac{\sqrt{57} - 7}{2 R_s^2} . \tag{3.16}$$

For this expression of C, the total GR mass $M = m(R_s)$ can then be computed from the usual definition (1.6) using the above density and we find

$$R_s = 2 n \left(\frac{\ell_P}{M_P}\right) \frac{M}{C_\rho} , \tag{3.17}$$

with $n \equiv \frac{56}{43-\sqrt{57}} \simeq 1.6$. This result, along with the choice [15]

$$K = \left(\frac{M_P}{M_G}\right)^2 \frac{\ell_G}{M_G} , \tag{3.18}$$

can be used in the constraint (3.8) to obtain $C_\rho = (M_G/M_P)^4$ and the ADM mass

$$\mathcal{M} = \frac{M^3}{M^2 + n_1 M_G^2} . \tag{3.19}$$

Moreover, the expression for the tidal charge (3.9) yields

$$q = \frac{\ell_G^2 M^2}{n \left(M^2 + n_1 M_G^2\right)} , \tag{3.20}$$

where we used $\sigma \simeq \ell_G^{-2}$ and

$$n_1 \equiv \frac{1}{2\,n^2}\left(\frac{5\,C\,R_s^2+1}{9\,C\,R_s^2+1}\right) = \frac{31653 - 1007\,\sqrt{57}}{175616} \simeq 0.14 \,. \qquad (3.21)$$

Upon inverting the relation (3.19) between $\mathcal{M}$ and M, one could finally express the tidal charge $q = q(\mathcal{M}, \sigma)$. The explicit expression of $M = M(\mathcal{M})$ is however rather cumbersome and we shall not display it here. Moreover, the expressions (3.17) and (3.19) satisfy the bounds in Eq. (3.10), and are therefore associated with a consistent boundary condition (3.8).

3.4 Black Hole Limit and Minimum Mass

The ADM mass (3.19) and tidal charge (3.20) do not explicitly depend on the radius R_s, and we can therefore assume they are valid in the (formal) limit $R_s \to 0$ (which one could more physically regard as $R_s \ll \ell_G$ [15]). Correspondingly, the exterior metric given by Eq. (3.11) becomes

$$e^{\nu_{(+)}} = 1 - \frac{2\,\ell_P\,M^3}{M_P\left(M^2 + n_1\,M_G^2\right)r}\left(1 + \frac{\ell_G^2\,M_P}{2\,n\,\ell_P\,M\,r}\right), \qquad (3.22)$$

which can be used to describe a black hole of "bare" mass M. Figure 3.1 shows the dimensionless ADM mass

$$\bar{\mathcal{M}} = \frac{\mathcal{M}}{M_G} = \frac{\bar{M}^3}{\bar{M}^2 + n_1} \simeq \frac{\bar{M}^3}{0.1 + \bar{M}^2} \,, \qquad (3.23)$$

and tidal charge

$$\bar{q} = \frac{q}{\ell_G^2} = \frac{\bar{M}^2}{n\,(n_1 + \bar{M}^2)} \simeq \frac{\bar{M}^2}{0.2 + 1.6\,\bar{M}^2} \,, \qquad (3.24)$$

as functions of the dimensionless proper mass $\bar{M} = M/M_G$. Figure 3.2 shows the dimensionless charge $\bar{q}$ as a function of the ADM mass $\bar{\mathcal{M}}$. Note that, for $M \ge M_G$, the ADM mass $\mathcal{M} \simeq M$, whereas the tidal charge saturates to a maximum

$$q_{max} \simeq 0.6\,\ell_G^2 \,, \qquad (3.25)$$

and is practically negligible for macroscopic black holes.

We can now derive an important result for microscopic black holes from Eq. (3.20). A black hole is usually considered (semi)classical if its horizon radius R_H is larger than the Compton wavelength $\lambda_M \simeq \ell_P\,M_P/M$ (see [16] and References therein). From Eq. (3.2), we obtain the horizon radius

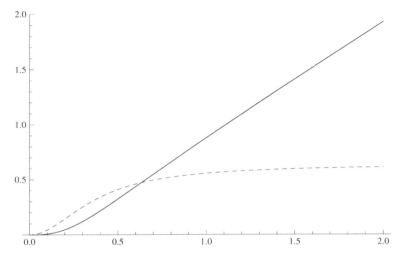

Fig. 3.1 Dimensionless ADM mass $\bar{\mathcal{M}}$ from Eq. (3.23) (solid line) and charge $\bar{q}$ from Eq. (3.24) (dashed line) *vs* the "bare" mass $\bar{M}$

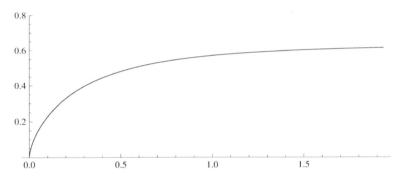

Fig. 3.2 Dimensionless charge $\bar{q}$ versus ADM mass $\bar{\mathcal{M}}$

$$R_{\mathrm{H}} = \frac{\ell_{\mathrm{P}}}{M_{\mathrm{P}}} \left(\mathcal{M} + \sqrt{\mathcal{M}^2 + q \, \frac{M_{\mathrm{P}}^2}{\ell_{\mathrm{P}}^2}} \right) , \tag{3.26}$$

and the classicality condition $R_{\mathrm{H}} \geq \lambda_M$ reads

$$\frac{M}{M_{\mathrm{P}}^2} \left(\mathcal{M} + \sqrt{\mathcal{M}^2 + q \, \frac{M_{\mathrm{P}}^2}{\ell_{\mathrm{P}}^2}} \right) \geq 1 . \tag{3.27}$$

We define the critical mass M_c as the value which saturates the above bound. In order to proceed, we shall expand for $M \sim \mathcal{M} \simeq M_{\mathrm{G}} \ll M_{\mathrm{P}}$, thus obtaining

$$\frac{R_{\text{H}}^2}{\lambda_M^2} \simeq \bar{M}^2 \bar{q} \simeq 1 \, , \tag{3.28}$$

or $\bar{M}^4 \simeq n \left(n_1 + \bar{M}^2\right)$, which yields

$$M_c \simeq 1.3 \, M_{\text{G}} \, , \tag{3.29}$$

or $\mathcal{M}_c \simeq 1.2 \, M_{\text{G}}$, from Eq. (3.23). This can be viewed as the minimum allowed mass for a semiclassical black hole in the BW.

3.4.1 More Linearly Charged Brane-World Black Holes

By following the same procedure, we can obtain more black hole configurations whose tidal charge and ADM mass are linearly related like in Eq. (3.9) by employing different expressions for the constants K and C_ρ. In particular, we shall always set $C_\rho = 1$ for simplicity.

The choice

$$K = \frac{\ell_{\text{P}}}{M_{\text{P}}} \, , \tag{3.30}$$

leads to the same asymptotic value (3.25) of the tidal charge, which therefore implies that q is practically negligible at all macroscopic scales. Moreover,

$$\mathcal{M} = \frac{M^3}{M^2 + \tilde{n} \, \frac{M_{\text{P}}^2}{M_{\text{G}}^2} \, M_{\text{P}}^2} \, , \tag{3.31}$$

and

$$q = \frac{\ell_{\text{G}}^2 \, M^2}{\tilde{n}_1 \, M^2 + \tilde{n}_2 \, \frac{M_{\text{P}}^2}{M_{\text{G}}^2} \, M_{\text{P}}^2} \, , \tag{3.32}$$

which shows that the ADM mass also becomes negligibly small for $M \sim M_{\text{G}}$,

$$\mathcal{M} \sim \left(\frac{M_{\text{G}}}{M_{\text{P}}}\right)^4 M \ll M \simeq M_{\text{G}} \, , \tag{3.33}$$

and analogously for the tidal charge

$$q \sim \ell_{\text{G}}^2 \left(\frac{M_{\text{G}}}{M_{\text{P}}}\right)^2 \left(\frac{M}{M_{\text{P}}}\right)^2 \ll \ell_{\text{G}}^2 \, . \tag{3.34}$$

Since it is $\mathcal{M}$ which measures the gravitational strength of a star in the weak field limit, one can view the above expressions as describing a screening of the star's gravitational field for very small GR mass. In fact, this screening effect seems somewhat similar to the vanishing of the ADM mass in the neutral shell model of Refs. [17, 18].

The choice

$$K = \left(\frac{\ell_{\mathrm{P}}}{M_{\mathrm{P}}}\right)\left(\frac{M_{\mathrm{G}}}{M_{\mathrm{P}}}\right)^{\alpha} \tag{3.35}$$

for $M \gg M_{\mathrm{G}}$ yields the asymptotic value

$$q_{\infty} = \frac{\ell_{\mathrm{G}}^2}{\tilde{n}_1}\left(\frac{M_{\mathrm{G}}}{M_{\mathrm{P}}}\right)^{\alpha}, \tag{3.36}$$

which can be very large or small depending on the sign of α. In particular, $\alpha > 0$ again implies a totally negligible asymptotic tidal charge. Further,

$$\mathcal{M} = \frac{M}{1 + \tilde{n}\left(\frac{M_{\mathrm{G}}}{M_{\mathrm{P}}}\right)^{\alpha-2}\left(\frac{M_{\mathrm{P}}}{M}\right)^2}, \tag{3.37}$$

and

$$q = \frac{(M_{\mathrm{G}}/M_{\mathrm{P}})^{\alpha}\,\ell_{\mathrm{G}}^2}{\tilde{n}_1 + \tilde{n}_2\left(\frac{M_{\mathrm{G}}}{M_{\mathrm{P}}}\right)^{\alpha-2}\left(\frac{M_{\mathrm{P}}}{M}\right)^2}, \tag{3.38}$$

which, for $M \sim M_{\mathrm{G}}$, respectively become

$$\mathcal{M} \sim \frac{M}{1 + \tilde{n}\left(\frac{M_{\mathrm{G}}}{M_{\mathrm{P}}}\right)^{\alpha-4}}, \tag{3.39}$$

and

$$q \sim \frac{(M_{\mathrm{G}}/M_{\mathrm{P}})^{\alpha}\,\ell_{\mathrm{G}}^2}{\tilde{n}_1 + \tilde{n}_2\left(\frac{M_{\mathrm{G}}}{M_{\mathrm{P}}}\right)^{\alpha-4}}. \tag{3.40}$$

For $M \simeq M_{\mathrm{G}}$, we then have

$$\mathcal{M} \simeq \begin{cases} \left(\frac{M_{\mathrm{G}}}{M_{\mathrm{P}}}\right)^{4-\alpha} M_{\mathrm{G}} \ll M_{\mathrm{G}} & \text{for } \alpha < 4 \\ M_{\mathrm{G}} & \text{for } \alpha \geq 4, \end{cases} \tag{3.41}$$

and

$$q \simeq \begin{cases} \ell_{\mathrm{G}}^2\left(\frac{M_{\mathrm{G}}}{M_{\mathrm{P}}}\right)^4 \ll \ell_{\mathrm{G}}^2 & \text{for } \alpha < 4 \\ \ell_{\mathrm{G}}^2\left(\frac{M_{\mathrm{G}}}{M_{\mathrm{P}}}\right)^{\alpha} \ll \ell_{\mathrm{G}}^2 & \text{for } \alpha \geq 4. \end{cases} \tag{3.42}$$

The first case ($\alpha < 4$) therefore generalises the kind of screening effect we described above, whereas the second case ($\alpha \geq 4$) would lead to a Schwarzschild-like exterior metric with ADM mass of the order of $M \simeq M_G$. However, as we noticed before, the tidal charge is always negligible for $\alpha > 0$ and the case $\alpha \geq 4$ is therefore of no physical interest, since it practically coincides with the standard Schwarzschild geometry for all GR masses $M \geq M_G$.

The last choice we consider is given by

$$K = \left(\frac{\ell_G}{M_G}\right)\left(\frac{M_G}{M_P}\right)^\beta ,$$
(3.43)

from which we obtain

$$q_\infty = \frac{\ell_G^2}{\tilde{n}_1}\left(\frac{M_G}{M_P}\right)^{\beta-2} ,$$
(3.44)

which coincides with the previous case for $\beta = \alpha + 2$. In fact, the ADM mass

$$\mathcal{M} = \frac{M}{1 + \tilde{n}\left(\frac{M_G}{M_P}\right)^{\beta-4}\left(\frac{M_P}{M}\right)^2} ,$$
(3.45)

and charge

$$q = \frac{(M_G/M_P)^{\beta-2}\ell_G^2}{\tilde{n}_1 + \tilde{n}_2\left(\frac{M_G}{M_P}\right)^{\beta-4}\left(\frac{M_P}{M}\right)^2} ,$$
(3.46)

in the limit for $M \sim M_G$, yield

$$\mathcal{M} \sim \frac{M}{1 + \tilde{n}\left(\frac{M_G}{M_P}\right)^{\beta-6}} ,$$
(3.47)

and

$$q \sim \frac{(M_G/M_P)^{\beta-2}\ell_G^2}{\tilde{n}_1 + \tilde{n}_2\left(\frac{M_G}{M_P}\right)^{\beta-6}} .$$
(3.48)

Again, in the limit $M \simeq M_G$, we obtain

$$\mathcal{M} \simeq \begin{cases} \left(\frac{M_G}{M_P}\right)^{\beta-n} M_G \ll M_G & \text{for} \quad \beta < 6 \\ M_G & \text{for} \quad \beta \geq 6 \end{cases}$$
(3.49)

and

$$q \simeq \begin{cases} \ell_{\mathrm{G}}^2 \left(\dfrac{M_{\mathrm{G}}}{M_{\mathrm{P}}} \right)^4 \ll \ell_{\mathrm{G}}^2 \quad \text{for} \quad \beta < 6 \\[4mm] \ell_{\mathrm{G}}^2 \left(\dfrac{M_{\mathrm{G}}}{M_{\mathrm{P}}} \right)^{\beta-2} \ll \ell_{\mathrm{G}}^2 \text{ for } \beta \geq 6 \,, \end{cases} \tag{3.50}$$

which reproduce the same behaviour as Eqs. (3.41) and (3.42). The case of $\beta \geq 2$ again leads to negligible tidal charge and we are just left with a large parameter range ($\beta < 2$) with gravitational screening. There is therefore hope that such a behaviour may represent a general feature of (some) BW metrics in the limit $M \simeq M_{\mathrm{G}}$.

3.4.2 Other Interior Geometries

The generality of the results presented so far can be assessed by considering other interior BW solutions. In particular, we shall here refer to the MGD of the non-uniform Heintzmann solution described in Sect. 2.1 and the MGD of the uniform Schwarzschild interior presented in Sect. 2.2.

For the non-uniform BW solution one finds the dimensionless ADM mass

$$\bar{\mathcal{M}} \simeq \frac{\bar{M}^3}{0.18 + \bar{M}^2} \tag{3.51}$$

and the tidal charge

$$\bar{q} \simeq \frac{\bar{M}^2}{0.24 + 1.3\,\bar{M}^2} \,. \tag{3.52}$$

Correspondingly, one has a minimum mass for semiclassical black holes given by

$$M_{\mathrm{c}} \simeq 1.2\,M_{\mathrm{G}} \,. \tag{3.53}$$

This value is slightly smaller than the one in Eq. (3.29), and one might also notice that the BW Heintzmann solution is more compact than the solution we employed to obtain that result.

On the other hand, the Schwarzschild BW solution yields

$$\bar{\mathcal{M}} \simeq \frac{\bar{M}^3}{0.1 + \bar{M}^2}\,s \tag{3.54}$$

and

$$\bar{q} \simeq \frac{\bar{M}^2}{0.18 + 1.8\,\bar{M}^2} \,. \tag{3.55}$$

Hence, the corresponding minimum mass for a semiclassical black hole would now be given by the somewhat larger

$$M_c \simeq 1.9 \, M_G \; . \tag{3.56}$$

The Schwarzschild interior solution represents the least compact distribution among the three solutions discussed in this Chapter, which suggests that less compact distributions correspond to greater critical mass M_c.

References

1. L. Randall, R. Sundrum, A large mass hierarchy from a small extra dimension. Phys. Rev. Lett. **83**, 3370 (1999)
2. L. Randall, R. Sundrum, An alternative to compactification. Phys. Rev. Lett. **83**, 4690 (1999)
3. M. Cavaglia, Black hole and brane production in TeV gravity: a review. Int. J. Mod. Phys. A **18**, 1843 (2003)
4. R. Casadio, L. Mazzacurati, Bulk shape of brane world black holes. Mod. Phys. Lett. A **18** 651–660 (2003). R. Casadio, O. Micu, Exploring the bulk of tidal charged micro-black holes. Phys. Rev. D **81** 104024 (2010)
5. R. Whisker, Braneworld Black Holes, Ph.D Thesis, arXiv:0810.1534 [gr-qc]. R. Gregory, *Braneworld Black Holes*. Lecture Notes in Physics, vol. 769 (2009), p. 259
6. P. Figueras, T. Wiseman, Gravity and large black holes in Randall-Sundrum II braneworlds. Phys. Rev. Lett. **107**, 081101 (2011)
7. D.-C. Dai, D. Stojkovic, Analytic solution for a static black hole in RSII model. Phys. Lett. B **704**, 354 (2011)
8. S. Abdolrahimi, C. Cattoen, D.N. Page, S. Yaghoobpour-Tari, Large Randall-Sundrum II black holes. Phys. Lett. B **720**, 405 (2013)
9. Marco Bruni, Cristiano Germani, Roy Maartens, Gravitational collapse on the brane: a no-go theorem. Phys. Rev. Lett. **87**, 231302 (2001)
10. M. Govender, N. Dadhich, Collapsing sphere on the brane radiates. Phys. Lett. B **538**, 233238 (2002)
11. S.W. Hawking, Black hole explosions. Nature **248**, 30 (1974); Particle creation by black holes. Comm. Math. Phys. **43**, 199 (1975)
12. J. Garriga, T. Tanaka, Gravity in the Randall-Sundrum brane world. Phys. Rev. Lett. **84**, 2778 (2000)
13. R. Casadio, C. Germani, Gravitational collapse and black hole evolution: do holographic black holes eventually 'anti-evaporate'? Prog. Theor. Phys. **114**, 23–56 (2005)
14. N. Dadhich, R. Maartens, P. Papadopoulos, V. Rezania, Black holes on the brane. Phys. Lett. B **487**, 1–6 (2000)
15. R. Casadio, J. Ovalle, Brane-world stars and (microscopic) black holes. Phys. Lett. B **715**, 251 (2012)
16. G.L. Alberghi, R. Casadio, O. Micu, A. Orlandi, Brane-world black holes and the scale of gravity. JHEP **1109**, 023 (2011)
17. R. Arnowitt, S. Deser, C.W. Misner, Finite self-energy of classical point particles. Phys. Rev. Lett. **4**, 375 (1960)
18. R. Casadio, R. Garattini, F. Scardigli, Point-like sources and the scale of quantum gravity. Phys. Lett. B **679**, 156 (2009)

Chapter 4
A Generalization of the Minimal Geometric Deformation

We have seen that the new terms in the effective four-dimensional Einstein field equations (1.14) originating from the bulk, which can be viewed as corrections to GR, might be the key to solve some open issues in gravity, like the dark matter problem. In this respect, from the phenomenological point of view, the search for solutions to the effective four-dimensional Einstein field equations for self-gravitating systems is of relevance, particularly in the case of vacuum solutions beyond the Schwarzschild metric. We already recalled that, due to the complexity of these effective equations, even in the simplest case of the RS BW, only a few candidate space-times of spherically symmetric self-gravitating systems are known exactly [1–6]. A useful guide in the search of such solutions is provided by the requirement that GR be recovered at low energies, which is at the foundation of the MGD approach as discussed in Chap. 1. Moreover, the RS model contains the brane tension σ as a free parameter, which allows one to control departures from GR by precisely setting the scale of high energy physics [7]. In light of all the results obtained from the application of the MGD approach [8–16], it appears natural to study possible extensions thereof, which is the topics of the present chapter.

In addition, the field equations in the bulk and the brane were shown to be consistent perturbatively [17–21]. In this context, the four-dimensional solution is naturally embedded in the five-dimensional bulk, with a black string-like object associated to the extended MGD procedure. This result has already been established in the case of the standard MGD technique [22]. On a fluid brane with variable tension, the event horizon of the black string along the extra dimension is a particular case of the bulk metric near the brane, being based merely upon the brane metric. Numerical techniques were employed to study Eötvös branes [23, 24] in the extended MGD framework, showing the embedding of the solution in the five-dimensional bulk. Such an embedding represents a largely employed method that has been previously investigated [17, 25, 26] and moreover applied to a variable tension brane [27],

J. Ovalle and R. Casadio, *Beyond Einstein Gravity*, SpringerBriefs in Physics,
https://doi.org/10.1007/978-3-030-39493-6_4

incorporating inflation [18, 19, 27], dark dust [20] and the realistic cases involving post-Newtonian approximations for the Casadio-Fabbri-Mazzacurati black string [2, 21].

4.1   The Exterior Geometry

For the sake of convenience, let us begin by recalling briefly the most relevant results of the MGD approach from Chap. 1. In particular, the radial component of the metric can be written as

$$
e^{-\lambda} = \mu(r) + \underbrace{e^{-I} \int^r \frac{e^I}{\frac{\nu'}{2} + \frac{2}{x}} \left[H(p, \rho, \nu) + \frac{1}{\sigma} \left(\rho^2 + 3\,\rho\,p \right) \right] dx + \beta\, e^{-I}}_{\text{geometric deformation}}
$$

$$
\equiv \mu(r) + f(r)\,,
\tag{4.1}
$$

where $\beta = \beta(\sigma)$ is a function of the brane tension σ, with the function (1.42) given by

$$
I(r, r_0) \equiv \int_{r_0}^r \frac{\nu'' + \nu'^2/2 + 2\nu'/x + 2/x^2}{\nu'/2 + 2/x}\, dx\,,
\tag{4.2}
$$

and $\mu = \mu(r)$ is the standard GR expression of the radial metric component. In particular, by assuming the space outside the star is empty, one has

$$
\mu(r) = \begin{cases} 1 - \dfrac{2\,M}{r}, & \text{for}\quad r > R_s \\[2ex] 1 - \dfrac{k^2}{r} \displaystyle\int_0^r x^2 \rho\, dx \equiv 1 - \dfrac{2\,m(r)}{r}, & \text{for}\quad r \le R_s \end{cases}
\tag{4.3}
$$

where R_s is the radius of the star and $m = m(r)$ denotes the standard GR interior mass function for $r < R_s$. The constant M, in general, depends on the brane tension σ and must take the value of the GR mass $M_0 = m(R_s)$ in the absence of extra-dimensional effects. A crucial role in the original MGD approach is played by the function H in Eq. (4.1), which vanishes for any time metric function $\nu = \nu_{GR}(r)$ that corresponds to a standard GR solution. The geometric deformation in Eq. (4.1) is correspondingly "minimal", in that it is simply given by contributions coming from the density and pressure of the source. One can thus start from a given GR solution $\nu = \nu_{GR}(r)$, then obtain the corresponding deformed BW radial metric function $\lambda = \lambda(r)$ by evaluating the integral in Eq. (4.1) with $H = 0$, and finally compute the BW time metric function $\nu = \nu(r)$ from the remaining field equations. It is worth noting that the correction $\nu - \nu_{GR}$ necessarily vanishes for $\sigma^{-1} \to 0$, and the starting GR solution is properly recovered in this low energy limit.

Let us remark that the parameter $\beta = \beta(\sigma)$ in Eq. (4.1) could also depend on the mass M of the self-gravitating system and must be zero in the GR limit. For interior solutions, the condition $\beta = 0$ has to be imposed in order to avoid singular solutions at the center $r = 0$. However, for vacuum solutions in the region $r > R_s$, where there is a Weyl fluid filling the space-time around the spherically symmetric stellar distribution, β does not need to be zero, hence there can be a geometric deformation associated to the Schwarzschild solution. Overall, β plays a crucial role in the search of BW exterior solutions, and in assessing their physical relevance. In fact, observational constraints on β from the classical tests of GR in the solar system were discussed in Sect. 2.6.

4.2 Extended Geometric Deformation

BW effects on spherically symmetric stellar systems have already been extensively studied (see, e.g. Refs. [3, 28–37]). In particular, the exterior (where $\rho = p = 0$) is in general filled with a Weyl fluid (arising from the bulk), which can affect stellar structures [14]. From the Einstein equations (1.27)–(1.29) for the exterior region $r > R_s$ we can identify the Weyl density $\mathcal{U}$, effective radial pressure $\tilde{p}_r$ and effective tangential pressure $\tilde{p}_t$, which are given by

$$\tilde{p}_r = \left(\frac{\mathcal{U}}{3} + \frac{2\mathcal{P}}{3}\right) , \tag{4.4}$$

$$\tilde{p}_t = \left(\frac{\mathcal{U}}{3} - \frac{\mathcal{P}}{3}\right) , \tag{4.5}$$

clearly illustrating the anisotropy of the exterior,

$$\Pi \equiv \tilde{p}_r - \tilde{p}_t = \mathcal{P} , \tag{4.6}$$

produced by extra-dimensional effects.

The original MGD approach allows us to study this region by generating modifications to the GR Schwarzschild metric with the correct low energy limit, as discussed in Chap. 1. The next natural step is thus to investigate a generalisation of the MGD for the exterior region filled with a Weyl fluid. This can be accomplished by considering, in addition to the geometric deformation (4.1) on the radial metric component, a geometric deformation on the time metric component [38], that is

$$\nu(r) = \nu_s + h(r) , \tag{4.7}$$

where ν_s is given by the Schwarzschild expression

$$e^{\nu_s} = 1 - \frac{2M}{r} \, , \tag{4.8}$$

and $h = h(r)$ is the time deformation produced by bulk gravitons, which should be proportional to σ^{-1} to also preserve the correct GR limit. Now, by using the expressions in Eq. (4.1) (with $H = \rho = p = 0$) and Eq. (4.7) in the vacuum equation

$$R_\mu^{\ \mu} = e^{-\lambda} \left(\nu'' + \frac{\nu'^2}{2} + 2\frac{\nu'}{r} + \frac{2}{r^2} \right) - \lambda' e^{-\lambda} \left(\frac{\nu'}{2} + \frac{2}{r} \right) - \frac{2}{r^2}$$
$$= 0 \, , \tag{4.9}$$

we obtain the following first order differential equation for the radial geometric deformation $f = f(r)$ in Eq. (4.1), in terms of the time geometric deformation $h = h(r)$,

$$\left(\frac{\nu'}{2} + \frac{2}{r} \right) f' + \left(\nu'' + \frac{\nu'^2}{2} + \frac{2\nu'}{r} + \frac{2}{r^2} \right) f + F(h) = 0 \, , \tag{4.10}$$

whose formal solution is given by

$$f(r) = e^{-I(r,R)} \left(\beta - \int_R^r \frac{e^{I(x,R)} \, F(h)}{\frac{\nu'}{2} + \frac{2}{x}} dx \right) . \tag{4.11}$$

The exponent $I = I(r, R)$ is again given by Eq. (4.2) and

$$F(h) = \mu' \frac{h'}{2} + \mu \left(h'' + \nu_s' h' + \frac{h'^2}{2} + 2\frac{h'}{r} \right) . \tag{4.12}$$

The exterior deformed radial metric component is thus expressed as

$$e^{-\lambda(r)} = 1 - \frac{2M}{r}$$
$$+ \underbrace{e^{-I(r,R)} \left(\beta - \int_R^r \frac{e^{I(x,R)} \, F(h)}{\frac{\nu'}{2} + \frac{2}{x}} dx \right)}_{\text{Geometric deformation}} . \tag{4.13}$$

We obtained a first important new result:

> **The temporal deformation $h(r)$ acts as a source for the radial deformation $f(r)$.**

Finally, by using the expressions in Eqs. (4.7) and (4.13) in the field Eqs. (1.27) and (1.28), the Weyl functions $\mathcal{U}$ and $\mathcal{P}$ are written in terms of the radial deformation f and time deformation h as

$$\frac{6\mathcal{U}}{k^2\sigma} = -\frac{f}{r^2} - \frac{f'}{r} , \tag{4.14}$$

$$\frac{12\mathcal{P}}{k^2\sigma} = \frac{4f}{r^2} + \frac{1}{r}\left[f' + 3f\frac{\mu'}{\mu} + 3\mu h' + 3fh' \right] . \tag{4.15}$$

To summarise, any given deformation $h = h(r)$ of the time component of the metric will induce a radial deformation $f = f(r)$ according to Eq. (4.13). In particular, a vanishing time deformation $h = 0$ will produce $F = 0$ and the corresponding geometric deformation will be again minimal. For the Schwarzschild geometry, this procedure yields the deformed exterior solution previously studied in Ref. [22] (we note that a constant h also produces $F = 0$, and corresponds to the rigid time rescaling $dT = e^{h/2} dt$).

4.3   Modified Exterior Solution

A more interesting case is of course provided by non-constant time deformations $h = h(r)$ such that $F(h) = 0$, which will still produce a "minimal" deformation (4.11). Let us therefore consider the non-linear differential equation

$$F(h) = 0 , \tag{4.16}$$

whose solution is given by the simple expression

$$e^{h/2} = a + \frac{b}{2M} \frac{1}{\sqrt{1 - 2M/r}} , \tag{4.17}$$

where the integration constants a and b can both be functions of the brane tension σ. Upon imposing that the space-time be asymptotically flat, that is

$$e^\nu \to 1 \quad \Rightarrow \quad h \to 0 , \tag{4.18}$$

for $r \to \infty$, one finds

$$a = 1 - \frac{b}{2M} , \tag{4.19}$$

and the time component of the metric has the final form

$$e^{\nu} = \left(1 - \frac{2M}{r}\right)\left[1 + \frac{b(\sigma)}{2M}\left(\frac{1}{\sqrt{1 - \frac{2M}{r}}} - 1\right)\right]^2, \tag{4.20}$$

for $r > 2M$, with the radial metric component given by

$$e^{-\lambda} = 1 - \frac{2M}{r} + \beta e^{-I}, \tag{4.21}$$

where I can be computed exactly but we omit it here for simplicity.

A particularly simple case is given when $\beta = 0$, which will produce no geometric deformation in the radial metric component, so that $\lambda = -\nu_s$ is exactly the Schwarzschild form in Eq. (4.8), and diverges for $r \to 2M$. However, due to the modified time component (4.20), this is now a real singularity, as can be seen by noting that the Kretschmann scalar $R_{\mu\nu\rho\sigma}R^{\mu\nu\rho\sigma}$ diverges at $r = 2M$. Moreover, this singularity is soft, since the higher order invariant $(\nabla_\kappa\nabla_\tau R_{\mu\nu\rho\sigma})(\nabla^\kappa\nabla^\tau R^{\mu\nu\rho\sigma})$, involving at least two derivatives of the curvature, further diverges at $r = 2M$. The above solution could therefore only represent the exterior geometry of a star with radius $R_s > 2M$. Since this solution has no deformation in its radial metric component, its dark radiation will be zero, $\mathcal{U} = 0$, as shown by Eq. (4.14). However, the Weyl function is given by

$$\frac{\mathcal{P}(r)}{\sigma} = -\frac{b\,M^2\,k^2}{2\left[2\sqrt{1 - \frac{2M}{r}} - b\,M\left(\sqrt{1 - \frac{2M}{r}} - 1\right)\right]r^3}, \tag{4.22}$$

which in fact diverges for $r \to 2M$.

Finally, Eq. (4.20) can be written for large r as

$$e^{\nu} \simeq 1 - \frac{2M - b}{r} - \frac{b\,(2M - b)}{4\,r^2}, \tag{4.23}$$

from which one can read off the gravitational mass

$$\mathcal{M} = M - \frac{b}{2} \tag{4.24}$$

and the tidal charge

$$Q = \frac{b\,(2M - b)}{4}, \tag{4.25}$$

in terms of which the real singularity is located at

$$r_c = 2\mathcal{M} \left(1 - \frac{Q}{\mathcal{M}^2} \right) . \tag{4.26}$$

It would be theoretically interesting to see if one can change the nature of the singularity at $r = 2M$ if $\beta = \beta(\sigma, M) \neq 0$. In fact, the Kretschmann scalar $R_{\mu\nu\rho\sigma} R^{\mu\nu\rho\sigma}$ might not diverge for $r \to 2M$ provided β satisfies a very complicated algebraic equation that depends on M and $b = b(\sigma)$. However, we showed in Sect. 2.6 that the classical tests in the solar system imply the very strong bound $|\beta| \leq (2.80 \pm 3.45) \times 10^{-11}$, and it is very unlikely that β meets this condition for arbitrary astrophysical masses M.

4.4 Five-Dimensional Solutions

Four-dimensional black hole solutions obtained by the MGD approach, or otherwise, can be extended into bulk solutions in order to analyse the corresponding five-dimensional causal structure. Let y denote the extra-dimensional Gaussian coordinate which parameterises geodesics emanating from the brane into the bulk, so that $dy = n_A\, dx^A$, where n^A are the components of a vector field normal to the brane.[1] The components g_{AB} of the five-dimensional bulk metric are related to the components $g_{\mu\nu} = g_{\mu\nu}(x^\alpha, 0)$ of the brane metric by

$$g_{AB}\, dx^A\, dx^B = g_{\mu\nu}(x^\alpha, y)\, dx^\mu\, dx^\nu + dy^2 , \tag{4.27}$$

where the brane is located at $y = 0$. In addition, the effective brane four-dimensional cosmological constant Λ, the bulk cosmological constant Λ_5, and the brane tension σ are fine-tuned by [25] $\Lambda = \frac{\kappa_5^2}{2} \left(\Lambda_5 + \frac{1}{6}\kappa_5^2 \sigma^2 \right)$, where $\kappa_5 = 8\pi G_5$ and G_5 denotes the five-dimensional gravitational constant. The brane extrinsic curvature, due to the junction conditions, reads

$$K_{\mu\nu} = -\frac{\kappa_5^2}{2} \left[T_{\mu\nu} + \frac{1}{3}(\sigma - T)\, g_{\mu\nu} \right] . \tag{4.28}$$

We assume that the bulk metric near the brane can be Taylor expanded along the extra dimension with respect to y [25–27],

$$g_{\mu\nu}(x^\alpha, y) = \sum_j g_{\mu\nu}^{(j)}(x^\alpha, 0)\, \frac{|y|^j}{j!} , \tag{4.29}$$

and one then finds that the expansion up to order $j = 4$ given in Ref. [27] is enough for a consistent analysis near the brane [27]. For the sake of conciseness, we only

---

[1] We denote five-dimensional coordinates with capital $A = 0, 1, 2, 3, 4$, whereas Greek indices are always used for coordinates on the four-dimensional brane.

display the metric up to the third order here, that is

$$g_{\mu\nu}(x^\alpha, y) = g_{\mu\nu}(x^\alpha) - \kappa_5^2 \left[T_{\mu\nu} + \frac{1}{3}(\sigma - T)g_{\mu\nu} \right] |y|$$

$$+ \left\{ \frac{\kappa_5^4}{2} \left[T_{\mu\alpha} T^\alpha{}_\nu + \frac{2}{3}(\sigma - T) T_{\mu\nu} \right] \right.$$

$$\left. - 2\,\mathcal{E}_{\mu\nu} + \frac{1}{3} \left[\frac{\kappa_5^4}{6}(\sigma - T)^2 - \Lambda_5 \right] g_{\mu\nu} \right\} \frac{y^2}{2!}$$

$$+ \left[2 K_{\mu\beta} K^\beta{}_\alpha K^\alpha{}_\nu - \mathcal{E}_{(\mu|\alpha} K^\alpha{}_{|\nu)} - \nabla^\rho B_{\rho(\mu\nu)} \right.$$

$$+ \frac{1}{6}\Lambda_5 g_{\mu\nu} K + K^{\alpha\beta} R_{\mu\alpha\nu\beta} - K\mathcal{E}_{\mu\nu} + 3\,K^\alpha{}_{(\mu}\mathcal{E}_{\nu)\alpha}$$

$$\left. + K_{\mu\alpha} K_{\nu\beta} K^{\alpha\beta} - K^2 K_{\mu\nu} \right] \frac{|y|^3}{3!} + \dots \tag{4.30}$$

where $g_{\mu\nu}(x^\alpha) = g_{\mu\nu}^{(0)}(x^\alpha, y = 0)$, $R_{\sigma\rho\mu\nu}$ are the components of the brane Riemann tensor—here $R_{\mu\nu}$ and R stand for the Ricci tensor and scalar curvature, respectively, and $B_{\tau\rho\sigma}$ represents the trace-free bulk Weyl tensor. When the brane has variable tension, the black string event horizon along the extra dimension is affected and the above series contains additional terms [27]. In particular, terms of order $|y|^3$ in the expansion (4.30), involving derivatives of the variable brane tension read [27]

$$g_{\mu\nu}^{(3)\,\text{extra}} = \frac{2}{3}\kappa_5^2 \left[(\nabla_{(\nu}\nabla_{\mu)}\sigma - \partial^\alpha \partial_\alpha \sigma)g_{\mu\nu} \right]. \tag{4.31}$$

Furthermore, terms of order y^4 can be obtained straightforwardly, although they are very cumbersome, and are given by Eq. (12) of Ref. [27]. In what follows, we shall only consider a time-dependent brane tension $\sigma = \sigma(t)$. More details about spherically symmetric or anisotropic branes can be found in Ref. [39].

Now, since the area of the five-dimensional horizon is determined by the metric component $g_{\theta\theta}(x^\alpha, y)$ [17, 25, 27], we consider the general spherically symmetric metric (1.4) with the coefficients (4.20) and (4.21). In fact, $g_{\theta\theta}(x^\alpha, y = 0) = R_H^2$ is precisely the black hole event horizon squared, where R_H denotes the coordinate singularity on the brane.

In BW models, we can take into account the huge variation of temperature in the Universe during the cosmological evolution. In this context, fluid membranes of Eötvös type [23] play a prominent role in phenomenological aspects. In fact, the so-called Eötvös law states that the tension of the fluid membrane does depend upon the temperature as $\sigma \sim (T_{\text{critical}} - T)$, where T_{critical} denotes the highest temperature compatible with the existence of the fluid membrane. Hence, the BW is originated in an early Universe that is extraordinarily hot, corresponding to $\sigma \approx 0$. The four-dimensional gravitational coupling constant $k^2 = 8\pi G_N$ and the brane tension σ are considered to be tiny in this context, which reinforces BW effects accordingly.

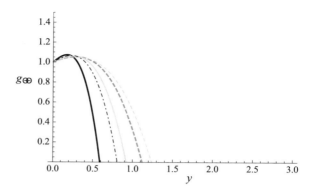

Fig. 4.1 Black string horizon radius squared $g_{\theta\theta}(r = 2M, y)$ along the extra dimension for $b(\sigma) = 1.9$ (solid black line), $b(\sigma) = 1.0$ (dash-dotted line), $b(\sigma) = 0.5$ (thick gray line), $b(\sigma) = 0.3$ (dashed gray line), $b(\sigma) = 0.1$ (dashed light gray line). Black hole mass $M = 1$ and $\beta \sim 1/\sigma$

In a suitable thermodynamic framework [24, 27], the time-dependent variable brane tension $\sigma(t) = 1 - 1/a(t)$, where $a(t)$ denotes the scale factor in a FLRW universe and is given in our particular case by $a(t) \sim 1 - \exp(\alpha t)$ [22, 27]. Additional phenomenological analysis of this model was done in Ref. [27]. Here we just consider $\Lambda_5 = 1 = \kappa_5$ and the brane tension with a lower bound of $\sigma \sim 4.39 \times 10^8$ MeV^4.

We can plot the black string event horizon $r = 2M$ for different values of the brane tension σ. Figure 4.1 shows how the shape of the black string event horizon along the extra dimension changes for different values of $\beta(\sigma) \sim \sigma^{-1}$. It is worth mentioning that there is a point of coordinate y_0 where the horizon satisfies $g_{\theta\theta}(r = 2M, y_0) = 0$. This point represents the intersection of the horizon with its axis of symmetry along the extra dimension. For instance, $y_0 \simeq 0.88$ when $\sigma = 0.5$, as one can see in Fig. 4.1.

The plot in Fig. 4.2 exhibits the time-dependent black string horizon along the extra dimension. As the cosmological time runs, the event horizon shrinks to a critical point along the extra dimension, whose constant time slices have been already depicted in Fig. 4.1 in a different range. These results are supported by the findings in Refs. [40, 42], and generalise those results to the extended MGD. From the perturbative analysis provided by the expansion (4.30), the black string horizon shrinks along the extra dimension. On the other hand, black strings are linearly unstable to long wavelength perturbations. If we perform the perturbation analysis according to Gregory and Laflamme [41], we can show that, although in the range $0.4 \leq y \leq 0.7$ the black string horizon associated to the extended MGD shrinks, in other regions along the extra dimension the event horizon's size can increase again, as for instance is shown in Ref. [22]. Hence the total area is shown to increase, consistently with thermodynamic principles. Early numerical analyses revealed that the black string fragments along the extra dimension, and forms five-dimensional black holes like those found in similar contexts [43]. However, a complete numerical analysis adapted to the MGD extended framework goes beyond the scope of this book.

Fig. 4.2 Time-dependent
black string event horizon
along the extra dimension,
for a variable tension brane

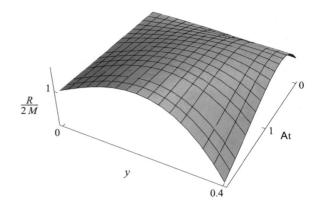

4.5   New Exact Solutions

We will next investigate a more general solution for Eq. (4.13) by considering a
time deformation $h = h(r)$ producing $F(h) \neq 0$. In particular, we shall consider
deformations of the Schwarzschild solution with the simple form

$$e^{\nu} = \left(1 - \frac{2M}{r}\right)^{1+K} , \tag{4.32}$$

corresponding to a time deformation

$$h = K \ln\left(1 - \frac{2M}{r}\right) . \tag{4.33}$$

We wish to emphasise that the parameter M in Eq. (4.32) is not the gravitational
mass $\mathcal{M}$ of the self-gravitating system. Indeed they only coincide when there is no
time deformation ($K = 0$). They are however related, as it is seen further below in
Equation (4.48).

Many exact configurations may be generated by changing the "deformation
parameter" K in Eq. (4.32). In fact, by replacing the metric component (4.32) with
generic parameter K into Eq. (4.13), the deformed radial metric component can be
computed exactly and reads

$$
\begin{aligned}
e^{-\lambda} = 1 - \frac{2M}{r} + \frac{(2 - 2M)^{1-2K}(-2M)^{K}}{r\left[(K-3)M + 2r\right]\left[(K-3)M\right]^{\frac{4K}{K-3}}} \\
\times \left[\beta r^{\frac{K(K+1)}{K-3}}\left(1 - \frac{r}{2M}\right)^{K}\left(1 + \frac{2r}{(K-3)M}\right)^{\frac{4K}{K-3}}\right. \\
\left. - (K-3)M r^{K}\left(1 - \frac{2M}{r}\right)\left[(K-3)M + 2r\right]^{\frac{4K}{K-3}} F(r, K)\right] , \tag{4.34}
\end{aligned}
$$

where

$$F(r, K) = \text{ApellF1}\left(a, b_1, b_2, c; \frac{r}{2M}, \frac{2r}{(3-K)M}\right) \tag{4.35}$$

is the Appell hypergeometric function of two variables with

$$a = \frac{K(K+1)}{3-K}; \quad b_1 = 1 - K; \quad b_2 = \frac{4K}{3-K}; \quad c = \frac{3+K^2}{3-K}. \tag{4.36}$$

The expression in Eq. (4.34) represents a kind of master equation associated with the general time deformation (4.33). A whole family of exact configurations may hence be generate by changing the deformation parameter K.

The simplest one is the MGD associated to the Schwarzschild solution, which corresponds to $K = 0$ and no time deformation. We recall that the radial metric component for this case reads

$$e^{-\lambda} = \left(1 - \frac{2M}{r}\right)\left(1 + \frac{\beta}{1 - \frac{3M}{2r}} \frac{\ell_0}{r}\right), \tag{4.37}$$

with the length scale

$$\ell_0 \equiv R_s \frac{(1 - \frac{3M}{2R_s})}{(1 - \frac{2M}{R_s})}, \tag{4.38}$$

and, of course, the Schwarzschild solution is recovered when $\beta = 0$.

For $K = 1$, Eqs. (4.32) and (4.13) respectively yield

$$e^{\nu} = 1 - \frac{4M}{r} + \frac{4M^2}{r^2} \tag{4.39}$$

and

$$e^{-\lambda} = 1 - \frac{(2M - c_1)}{r} + \frac{(2M^2 - c_1 M)}{r^2}, \tag{4.40}$$

where

$$c_1 \equiv \frac{R_s}{1 - M/R_s} \beta \tag{4.41}$$

and $r_0 = R_s$, the radius of the stellar distribution which was used to evaluate the integral in Eq. (4.2).

In order to obtain the correct asymptotic Schwarzschild behaviour

$$e^{-\lambda} \sim 1 - \frac{2M}{r} + \mathcal{O}(r^{-2}) \tag{4.42}$$

the constant c_1 must satisfy

$$c_1 = -2\,M\,. \tag{4.43}$$

Consequently, the expressions in Eqs. (4.39) and (4.40) will reproduce the tidally charged solution given by [1]

$$e^\nu = e^{-\lambda} = 1 - \frac{2\,M}{r} + \frac{Q}{r^2}\,, \tag{4.44}$$

where the ADM mass $\mathcal{M} = 2\,M$ and the tidal charge $Q = 4\,M^2$. The black hole solution in Eq. (4.44) corresponds to an extremal black hole with degenerate horizons

$$r_h = M \tag{4.45}$$

which lies inside the Schwarzschild radius. Hence, according to this solution, extra-dimensional effects weaken the gravitational field.

4.5.1   Outer Solutions

We can now show two examples of exact solutions for specific values of K which can be used to describe the region outside a compact source.

For $K = 2$, Eqs. (4.32) and (4.13) lead to

$$e^\nu = 1 - \frac{2\,M}{r} + \frac{Q}{r^2} - \frac{2\,M\,Q}{9\,r^3}\,, \tag{4.46}$$

and

$$
\begin{aligned}
e^{-\lambda} = {} & \left(1 - \frac{2\,M}{3\,r}\right)^{-1} \\
& \times \left[\frac{128\,c_2}{r}\left(1 - \frac{M}{6\,r}\right)^7 + \frac{5}{224}\left(\frac{Q}{12r^2}\right)^4 - \frac{5}{16}\frac{M}{3r}\left(\frac{Q}{12r^2}\right)^3 + \frac{5}{6}\left(\frac{Q}{12r^2}\right)^3 \right. \\
& \left. - \frac{25}{4}\frac{M}{3r}\left(\frac{Q}{12r^2}\right)^2 + \frac{25}{2}\left(\frac{Q}{12r^2}\right)^2 - \frac{5}{12}\frac{M}{r}\frac{Q}{r^2} + \frac{10}{12}\frac{Q}{r^2} - \frac{4}{3}\frac{M}{r} + 1 \right],
\end{aligned}
\tag{4.47}
$$

where

$$\mathcal{M} = 3\,M; \quad Q = 12\,M^2; \quad c_2 \equiv \frac{(1 - 2\,M/R)}{(2 - M/R)^7}\,\beta\,. \tag{4.48}$$

From the Schwarzschild limit in Eq. (4.42), we then obtain

Fig. 4.3 Case $K = 2$ with $\mathcal{M} = 1$. Behaviour of g_{tt} and g_{rr}^{-1}: graphs show two zeros and a singular point for g_{rr}^{-1}, the interior $r_i \simeq 0.095$, the middle $r_c = 2/3$, and the exterior $r_e \simeq 1.124$. The horizon is shifted inside the Schwarzschild radius $(r_s = 2\mathcal{M})$

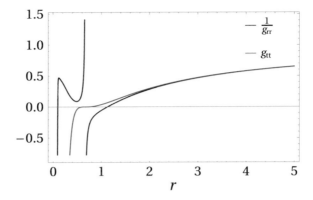

$$c_2 = -\frac{M}{32}, \tag{4.49}$$

which therefore fixes the constant β. This solution displays two zeros of g_{rr}^{-1}, namely $r = r_i$ and r_e, and a surface $r = r_c$ where g_{rr}^{-1} diverges (and $g_{tt} = 0$), all shown in Fig. 4.3. These three surfaces separate the space-time in four regions, namely

- $0 < r < r_i$
- $r_i < r < r_c$
- $r_c < r < r_e$
- $r > r_e$.

An exterior observer at $r > r_e > r_c$ will never see the singularity at $r = r_c$, as it is hidden behind the outer horizon $r = r_e$. Nonetheless, we prefer to consider this solution as a candidate exterior for a self-gravitating distribution of matter with radius $R_s > r_c$, thus excluding the singular surface at $r = r_c$. Finally, since the exterior horizon lies inside the Schwarzschild radius (obtained by setting the tidal charge $Q = 0$), $r_e < r_s = 2\mathcal{M}$, this solution also indicates that extra-dimensional effects weaken the gravitational field.

Both Weyl functions $\mathcal{U}$ and $\mathcal{P}$ are shown in Fig. 4.4. We see that $\mathcal{U}$ is always positive, which indicates a negative radial deformation, according to Eq. (4.14), and diverges at the singular surface $r = r_c$. On the other hand, $\mathcal{P}$ is always negative, showing thus a negative "pressure" around the stellar distribution consequently to extra-dimensional effects. This function also diverges at the singular radius $r = r_c$.

Another interesting example is obtained with $K = 4$, which represents a solution with higher-order deformation in terms of a tidal charge Q, namely

$$e^\nu = 1 - \frac{2\mathcal{M}}{r} + \frac{Q}{r^2} - \frac{2}{5}\frac{\mathcal{M}Q}{r^3} + \frac{Q^2}{20\,r^4} - \frac{\mathcal{M}Q^2}{250\,r^5}, \tag{4.50}$$

where $\mathcal{M} = 5\,M$ and $Q = 40\,M^2$. Similarly the radial metric component can be expressed in exact form, but it is so large we do not display it here. This solution

Fig. 4.4 Case $K = 2$
with $\mathcal{M} = 1$. Behaviour of
the Weyl function $\mathcal{U}$ and $\mathcal{P}$.
The scalar $\mathcal{U}$ increases on
approaching the source and
reaches a maximum inside
the horizon, then it diverges
towards the critical radius r_c.
The opposite behaviour is
seen for $\mathcal{P}$

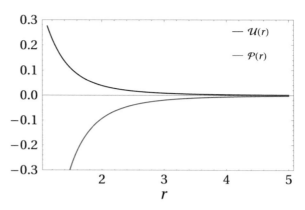

Fig. 4.5 Case $K = 4$ with
$\mathcal{M} = 1$. Behaviour of g_{tt}
and g_{rr}^{-1}. We see a zero and a
singular point for g_{rr}^{-1}. The
singularity at $r_c = 2/5$ is
hidden behind the horizon
$r_h \simeq 1.131$. The black hole
horizon is shifted inside the
Schwarzschild radius
$(r_s = 2\mathcal{M})$

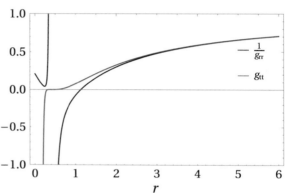

displays a zero of g_{rr}^{-1}, namely a horizon $r = r_h$, and a surface $r = r_c$ where g_{rr}^{-1}
diverges (and $g_{tt} = 0$), all shown in Fig. 4.5. These surfaces separate the space-time
in three regions, namely

- $0 < r < r_c$
- $r_c < r < r_h$
- $r > r_h$.

We emphasise that an exterior observer at $r > r_h > r_c$ will never see the singularity
at $r = r_c$, since it is hidden behind the horizon $r = r_h$. Finally, since the horizon lies
inside the Schwarzschild radius, $r_h < r_s = 2\mathcal{M}$, this solution once more indicates
that extra-dimensional effects weaken the gravitational field of compact sources.

The Weyl functions $\mathcal{U}$ and $\mathcal{P}$ are shown in Fig. 4.6. Both are positive and with
$\mathcal{P} > \mathcal{U}$. This indicates a negative radial deformation and also a positive temporal
deformation, according to Eqs. (4.14) and (4.15), which diverges at the singular
surface $r = r_c$.

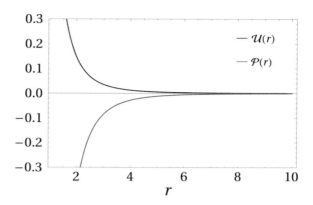

Fig. 4.6 Case $K = 4$ with $\mathcal{M} = 1$. Behaviour of the Weyl function $\mathcal{U}$ and $\mathcal{P}$. The scalar $\mathcal{U}$ increases towards the source and reaches a maximum inside the horizon, then it diverges towards the critical value r_c. The opposite behaviour is seen for $\mathcal{P}$

— $\mathcal{U}(r)$

— $\mathcal{P}(r)$

4.5.2 The Interior

So far we have successfully developed an extension of the MGD for the region $r > R_s$ surrounding a self-gravitating distribution. The next logical step is to consider this extension inside the distribution with $r < R_s$, in order to investigate the effects of bulk gravitons on physical variables such as the density and pressure. A critical point regarding the implementation of the MGD is the conservation equation (1.30), which contrary to GR, is not a linear combination of the field Eqs. (1.27)–(1.29). We know that in the MGD approach any chosen GR solution will automatically satisfy the conservation equation (1.30). However, when the time deformation in Eq. (4.7) is considered, the conservation equation (1.30) becomes

$$p' = -\frac{\nu_0'}{2}(\rho + p) - \frac{h'}{2}(\rho + p) \tag{4.51}$$

where ν_0 is the GR solution of the temporal metric component. The expression in Eq. (4.51) then yields

$$0 = -\frac{h'}{2}(\rho + p) \tag{4.52}$$

and therefore only a constant time deformation h, or equivalently a rigid time transformation $dT = e^{h/2} dt$, may be implemented inside a self-gravitating system formed by a perfect fluid. The only way to overcome this problem, in the context of the MGD, is by considering a more complex interior structure, like a fluid with anisotropic pressure, which leads to a more complex form of the conservation equation. Furthermore, an exchange of energy between the bulk and the brane could be useful to implement the extension of the MGD inside a stellar distribution. However, a successful extension for the interior of a self-gravitating system can be developed by using the so-called Gravitational Decoupling [44, 45], which is based in the MGD approach, briefly developed in the last chapter of this book.

References

1. N. Dadhich, R. Maartens, P. Papadopoulos, V. Rezania, Black holes on the brane. Phys. Lett. B **487**, 1–6 (2000)
2. R. Casadio, A. Fabbri, L. Mazzacurati, New black holes in the brane world? Phys. Rev. D **65**, 084040 (2002)
3. P. Figueras, T. Wiseman, Gravity and large black holes in Randall-Sundrum II braneworlds. Phys. Rev. Lett. **107**, 081101 (2011)
4. D.-C. Dai, D. Stojkovic, Analytic solution for a static black hole in RSII model. Phys. Lett. B **704**, 354 (2011)
5. S. Abdolrahimi, C. Cattoen, D.N. Page, S. Yaghoobpour-Tari, Large Randall-Sundrum II black holes. Phys. Lett. B **720**, 405 (2013)
6. T. Wiseman, Relativistic stars in Randall-Sundrum gravity. Phys. Rev. D **65**, 124007 (2002)
7. J. Ovalle, Searching exact solutions for compact stars in braneworld: a conjecture. Mod. Phys. Lett. A **23**, 3247 (2008)
8. J. Ovalle, Non-uniform braneworld stars: an exact solution. Int. J. Mod. Phys. D **18**, 837 (2009)
9. J. Ovalle, F. Linares, Tolman IV solution in the Randall-Sundrum braneworld. Phys. Rev. D **88**, 104026 (2013)
10. J. Ovalle, Searching exact solutions for compact stars in braneworld: a conjecture. Mod. Phys. Lett. A **23**, 3247 (2008)
11. J. Ovalle, Braneworld stars: anisotropy minimally projected onto the brane, in *Gravitation and Astrophysics* (World Scientific Press, Singapore, 2010), pp. 173–182
12. R. Casadio, J. Ovalle, Brane-world stars and (microscopic) black holes. Phys. Lett. B **715**, 251 (2012)
13. R. Casadio, J. Ovalle, Brane-world stars from minimal geometric deformation, and black holes. Gen. Relat. Grav. **46**, 1669 (2014)
14. J. Ovalle, F. Linares, A. Pasqua, A. Sotomayor, The role of exterior Weyl fluids on compact stellar structures in Randall-Sundrum gravity. Class. Quant. Grav. **30**, 175019 (2013)
15. L.Á. Gergely, Friedmann branes with variable tension. Phys. Rev. D **78**, 084006 (2003)
16. J. Ovalle, L.A. Gergely, R. Casadio, Brane-world stars with solid crust and vacuum exterior. Class. Quant. Grav. **32**, 045015 (2015)
17. R. Casadio, B. Harms, Black hole evaporation and compact extra dimensions. Phys. Rev. D **64**, 024016 (2001)
18. D. Bazeia, J.M. Hoff da SIlva, R. da Rocha: Black holes in realistic branes: black string-like objects?. Phys. Lett. B **721**, 306 (2013)
19. D. Bazeia, J.M. Hoff da Silva, R. da Rocha, Regular bulk solutions and black strings from dynamical braneworlds with variable tension. Phys. Rev. D **90**, 047902 (2014)
20. A. Herrera-Aguilar, A.M. Kuerten, R. da Rocha, Regular bulk solutions in brane-worlds with inhomogeneous dust and generalized dark radiation. Adv. High Energy Phys. **2015**, 359268 (2015)
21. R. da Rocha, A. Piloyan, A.M. Kuerten, Casadio-Fabbri-Mazzacurati black strings and braneworld-induced quasars luminosity corrections. Class. Quant. Grav. **30**, 045014 (2013)
22. R. Casadio, J. Ovalle, R. da Rocha, Black strings from minimal geometric deformation in a variable tension brane-world class. Quant. Grav. **30**, 175019 (2014)
23. R. Eötvös, Ueber den Zusammenhang der Oberflächenspannung der Flüssigkeiten mit ihrem Molecularvolumen. Ann. Phys. **263**, 448 (1886)
24. L.A. Gergely, Eötvös branes. Phys. Rev. D **79**, 086007 (2009)
25. R. Maartens, K. Koyama, Brane-world gravity. Living Rev. Rel. **13**, 5 (2010)
26. R. Casadio, C. Germani, Gravitational collapse and black hole evolution: do holographic black holes eventually 'anti-evaporate'? Prog. Theor. Phys. **114**, 23 (2005)
27. R. da Rocha, J.M. Hoff da Silva, Black string corrections in variable tension braneworld scenarios. Phys. Rev. D **85**, 046009 (2012)
28. R. Casadio, J. Ovalle, R. da Rocha, Classical tests of GR: brane-world sun from minimal geometric deformation. Europhys. Lett. **110**, 40003 (2015)

29. P. Kanti, N. Pappas, K. Zuleta, On the localisation of 4-dimensional brane-world black holes. Class. Quant. Grav. **30**, 235017 (2013)
30. T. Harko, M.J. Lake, Null fluid collapse in brane world models. Phys. Rev. D **89**, 064038 (2014)
31. L.B. Castro, M.D. Alloy, D.P. Menezes, Mass radius relation of compact stars in the braneworld. JCAP08 047 (2014)
32. M.A. García-Aspeitia, L.A. Ureña-López, Stellar stability in brane-worlds revisited. Class. Quant. Grav. **32**, 025014 (2015)
33. F.X. Linares, M.A. Garcia-Aspeitia, L. Arturo Ure-Lopez, Stellar models in brane worlds. Phys. Rev. D **92**, 024037 (2015)
34. Z. Stuchlik, J. Hladik, M. Urbanec, Neutrino trapping in braneworld extremely compact stars. Gen. Relativ. Gravit. **43**, 3163–3190 (2011)
35. J. Hladik, Z. Stuchlik, Photon and neutrino redshift in the field of braneworld compact stars. JCAP **07**, 012 (2011)
36. G. Kofinas, E. Papantonopoulos, I. Pappa, Spherically symmetric braneworld solutions with R_4 term in the bulk. Phys. Rev. D **66**, 104014 (2002)
37. S. Creek, R. Gregory, P. Kanti, B. Mistry, Braneworld stars and black holes. Class. Quant. Grav. **23**, 6633 (2006)
38. R. Casadio, J. Ovalle, R. Rocha, The minimal geometric deformation approach extended. Class. Quant. Grav. **32**(21), 215020 (2015)
39. A.E. Bernardini, R.T. Cavalcanti, R. da Rocha, Spherically symmetric thick branes cosmological evolution. Gen. Relativ. Gravit. **47**, 1840 (2015)
40. G.T. Horowitz, K. Maeda, Fate of the black string instability. Phys. Rev. Lett. **87**, 131301 (2001)
41. R. Gregory, R. Laflamme, Black strings and p-branes are unstable. Phys. Rev. Lett. **70**, 2837 (1993)
42. B. Kol, The phase transition between caged black holes and black strings: a review. Phys. Rep. **422**, 119 (2006)
43. L. Lehner, F. Pretorius, Black strings, low viscosity fluids, and violation of cosmic censorship. Phys. Rev. Lett. **105**, 101102 (2010)
44. J. Ovalle, Decoupling gravitational sources in GR: from perfect to anisotropic fluids. Phys. Rev. D **95**, 104019 (2017)
45. J. Ovalle, Decoupling gravitational sources in GR: the extended case. Phys. Lett. B **787** (2019)

Chapter 5
Gravitational Decoupling

Throughout this book, we have described in detail how to implement the MGD formalism in order to extend GR solutions for perfect fluids and the vacuum into the BW domain. We have tested the effectiveness and usefulness of this scheme in the extra-dimensional scenario, but the fundamental reason why it works the way it does was not yet explained. This is precisely the objective of this final chapter. We will see that the MGD scheme has to do with the decoupling of the gravitational sources in the Einstein field equations [1, 2], something which is not trivial, given the complexity and non-linearity of that system of equations.

We will show that this approach can indeed be applied to any pair of gravitational sources, regardless of their nature or origin, showing thus that under the MGD lies a powerful and direct way of dealing with the Einstein field equations [3–43] beyond the extra-dimensional context of the BW.

5.1 MGD Decoupling for Two Sources

Let us start with the following question: is it possible to solve the Einstein field equations containing two gravitational sources, represented by the energy-momentum tensors $T_{\mu\nu}$ and $\theta_{\mu\nu}$ [44], if we can solve some (suitable) field equations for each gravitational source individually? In other world, instead of finding the metric $g_{\mu\nu}$ by solving directly

$$G_{\mu\nu} = k^2 \left(T_{\mu\nu} + \theta_{\mu\nu} \right) \equiv k^2 \, \tilde{T}_{\mu\nu} , \tag{5.1}$$

suppose we can find the solution $\hat{g}_{\mu\nu}$ of an equation

$$\hat{G}_{\mu\nu} = k^2 \, T_{\mu\nu} \tag{5.2}$$

© The Author(s), under exclusive license to Springer Nature Switzerland AG 2020
J. Ovalle and R. Casadio, *Beyond Einstein Gravity*, SpringerBriefs in Physics,
https://doi.org/10.1007/978-3-030-39493-6_5

and then the solution $g^*_{\mu\nu}$ of another equation

$$G^*_{\mu\nu} = k^2\,\theta_{\mu\nu}\,. \tag{5.3}$$

such that we finally obtain the metric $g_{\mu\nu}$ which solves Eq. (5.1) by some *simple* combination of the two metrics $\hat{g}_{\mu\nu}$ and $g^*_{\mu\nu}$. Obviously, if this procedure worked, it would introduce an unprecedented simplification. In fact, for static spherically symmetric systems, such a "simple combination" of the two solutions $\hat{g}_{\mu\nu}$ and $g^*_{\mu\nu}$ can be found and is precisely the MGD of the radial component of the metric generically represented by the expression in Eq. (1.49). We will now show that the decoupling of the gravitational sources, hinted in Eqs. (5.1)–(5.3), can be done in a simple and direct way, at least for the spherically symmetric and static systems, thus opening a range of new possibilities in the search for solutions to the Einstein field equations.

Let us start from the standard Einstein field equations (5.1). Since the Einstein tensor satisfies the Bianchi identity, the total source must satisfy the conservation equation

$$\nabla_\mu \tilde{T}^{\mu\nu} = 0\,. \tag{5.4}$$

In the static and spherical symmetric case, described by a metric in the usual form (1.4), the above Einstein equations explicitly read

$$k^2\left(T_0{}^0 + \theta_0^0\right) = \frac{1}{r^2} - e^{-\lambda}\left(\frac{1}{r^2} - \frac{\lambda'}{r}\right) \tag{5.5}$$

$$k^2\left(T_1{}^1 + \theta_1^1\right) = \frac{1}{r^2} - e^{-\lambda}\left(\frac{1}{r^2} + \frac{\nu'}{r}\right) \tag{5.6}$$

$$k^2\left(T_2{}^2 + \theta_2^2\right) = -\frac{e^{-\lambda}}{4}\left(2\nu'' + \nu'^2 - \lambda'\nu' + 2\frac{\nu' - \lambda'}{r}\right)\,, \tag{5.7}$$

where $f' \equiv \partial_r f$ and $\tilde{T}_3{}^3 = \tilde{T}_2{}^2$ due to the spherical symmetry. The conservation equation (5.4) is a linear combination of Eqs. (5.5)–(5.7), and yields

$$\left(\tilde{T}_1{}^1\right)' - \frac{\nu'}{2}\left(\tilde{T}_0{}^0 - \tilde{T}_1{}^1\right) - \frac{2}{r}\left(\tilde{T}_2{}^2 - \tilde{T}_1{}^1\right) = 0\,, \tag{5.8}$$

which in terms of the two sources in Eq. (5.1) read

$$\left(T_1{}^1\right)' - \frac{\nu'}{2}\left(T_0{}^0 - T_1{}^1\right) - \frac{2}{r}\left(T_2{}^2 - T_1{}^1\right)$$
$$+ \left[\left(\theta_1^1\right)' - \frac{\nu'}{2}\alpha\left(\theta_0^0 - \theta_1^1\right) - \frac{2\alpha}{r}\left(\theta_2^2 - \theta_1^1\right)\right] = 0\,. \tag{5.9}$$

By simple inspection, we can identify in Eqs. (5.5)–(5.7) an effective density $\tilde{\rho}$, an effective radial pressure $\tilde{p}_r$ and an effective tangential pressure $\tilde{p}_t$ respectively given by

$$\tilde{\rho} = T_0{}^0 + \theta_0{}^0 \; ; \quad \tilde{p}_r = -T_1{}^1 - \theta_1{}^1 \; ; \quad \tilde{p}_t = -T_2{}^2 - \theta_2{}^2 \; , \tag{5.10}$$

for which the anisotropy is given by

$$\Pi \equiv \tilde{p}_t - \tilde{p}_r \; . \tag{5.11}$$

The system of Eqs. (5.5)–(5.7) may therefore be formally treated as a fluid with a possibly anisotropic pressure [45, 46].

The MGD decoupling can now be applied. We start by considering a solution to the Eq. (5.2) for the source $T_{\mu\nu}$ alone [that is Eqs. (5.5)–(5.8) with $\theta_{\mu\nu} = 0$], which we can write as

$$ds^2 = e^{\xi(r)} dt^2 - \frac{dr^2}{\mu(r)} - r^2 \left(d\theta^2 + \sin^2 \theta \, d\phi^2 \right) , \tag{5.12}$$

where

$$\mu(r) \equiv 1 - \frac{k^2}{r} \int_0^r x^2 \, T_0{}^0(x) \, dx = 1 - \frac{2 \, m(r)}{r} \tag{5.13}$$

is the standard GR expression containing the Misner-Sharp mass function $m(r)$.

The effects of the source $\theta_{\mu\nu}$ on $T_{\mu\nu}$ can then be encoded in the MGD undergone solely by the radial component of the perfect fluid geometry in Eq. (5.12). Namely, the general solution is given by Eq. (1.49) with $\nu(r) = \xi(r)$ and

$$e^{-\lambda(r)} = \mu(r) + \alpha \, f^*(r) , \tag{5.14}$$

where f^* is the (minimal) geometric deformation due to the effects of the source $\theta_{\mu\nu}$ and α a constant that helps to keep track of these effects.

Now let us plug the expression (5.14) into the Einstein equations (5.5)–(5.7). This system of equations is thus separated into two sub-systems: (i) the first one is given by the standard Einstein field equations (5.2) with the energy-momentum tensor $T_{\mu\nu}$, whose solution is given by Eq. (5.12),

$$k^2 \, T_0{}^0 = \frac{1}{r^2} - \frac{\mu}{r^2} - \frac{\mu'}{r} \tag{5.15}$$

$$k^2 \, T_1{}^1 = \frac{1}{r^2} - \mu \left(\frac{1}{r^2} + \frac{\xi'}{r} \right) \tag{5.16}$$

$$k^2 \, T_2{}^2 = -\frac{\mu}{4} \left(2 \, \xi'' + \xi'^2 + \frac{2 \, \xi'}{r} \right) - \frac{\mu'}{4} \left(\xi' + \frac{2}{r} \right) , \tag{5.17}$$

and its respective conservation equation

$$\left(T_1{}^1\right)' - \frac{\xi'}{2}\left(T_0{}^0 - T_1{}^1\right) - \frac{2}{r}\left(T_2{}^2 - T_1{}^1\right) = 0\,, \tag{5.18}$$

and (ii) the second sub-system is given by the "quasi-Einstein" field equations (5.3) for the source $\theta_{\mu\nu}$, which reads

$$k^2\,\theta_0^0 = -\frac{\alpha\,f^*}{r^2} - \frac{\alpha\,f^{*\prime}}{r} \tag{5.19}$$

$$k^2\,\theta_1^1 = -\alpha\,f^*\left(\frac{1}{r^2} + \frac{\xi'}{r}\right) \tag{5.20}$$

$$k^2\,\theta_2^2 = -\frac{\alpha}{4}\left[f^*\left(2\,\xi'' + \xi'^2 + 2\frac{\xi'}{r}\right) + f^{*\prime}\left(\xi' + \frac{2}{r}\right)\right]\,, \tag{5.21}$$

and its conservation equation

$$\left(\theta_1^1\right)' - \frac{\xi'}{2}\left(\theta_0^0 - \theta_1^1\right) - \frac{2}{r}\left(\theta_2^2 - \theta_1^1\right) = 0\,. \tag{5.22}$$

From the expressions (5.9), (5.18) and (5.22) we see that there is no direct exchange of energy between the sources $T_{\mu\nu}$ and $\theta_{\mu\nu}$ and therefore their interaction is purely gravitational. We conclude that we have successfully decoupled the Einstein equations (5.5)–(5.7) in two systems, namely: (i) the Einstein field equations for a source $T_{\mu\nu}$ displayed in Eqs. (5.15)–(5.17) which can be used to determine the pair of metric functions $\{\xi,\ \mu\}$, and (ii) the field equations for the source $\theta_{\mu\nu}$ shown in Eqs. (5.19)–(5.21) from which one can determine the MGD f^* for the radial metric function (let us recall that there is no deformation on the temporal metric component, hence $\nu = \xi$).

We emphasise that, to our knowledge, the MGD decoupling represents the first simple and systematic approach to decouple gravitational sources in GR. The two main applications of this approach are the following [1]:

- *Extending simple solutions into more complex domains.* We can start from a simple energy-momentum tensor $T_{\mu\nu}$ and add to it more complex gravitational sources. The starting source $T_{\mu\nu}$ could be as simple as we wish, including the vacuum indeed, to which we can add a first new source, say

$$T_{\mu\nu} \mapsto \tilde{T}_{\mu\nu}^{(1)} = T_{\mu\nu} + T_{\mu\nu}^{(1)}\,. \tag{5.23}$$

We can then repeat the process with more sources. In this way, we can extend solutions of the Einstein equations associated with the simplest gravitational source $T_{\mu\nu}$ into the domain of more complex forms of gravitational sources $T_{\mu\nu} = \tilde{T}_{\mu\nu}^{(n)}$, step by step and systematically.

- *Deconstructing a complex gravitational source.* In order to find a solution to the Einstein equations with a complex energy-momentum tensor $\tilde{T}_{\mu\nu}$, we can split it into simpler components, say $T_{\mu\nu}$ and $T^{(i)}_{\mu\nu}$, and solve the Einstein equations for each one of these parts. Hence, we will find as many solutions as are the contributions $T^{(i)}_{\mu\nu}$ in $\tilde{T}_{\mu\nu}$. Finally, by a straightforward combination of all these solutions, we will obtain the solution to the Einstein equations associated with the original energy-momentum tensor $\tilde{T}_{\mu\nu}$.

Since the Einstein field equations are non-linear, the MGD decoupling represents a powerful tool in the search and analysis of solutions, especially when we deal with situations beyond trivial cases.

In the next section, we shall apply the MGD decoupling to the effective Einstein equations in the BW.

5.2 Gravitational Decoupling in the Brane-World

Let us implement the gravitational decoupling to the BW system of field equations (1.27)–(1.29). If we compare the effective energy-momentum tensor in Eq. (1.16) with the right hand side of Eq. (5.1), we straightforwardly find that the energy-momentum tensor $T_{\mu\nu}$ is the purely four-dimensional matter contribution, which we usually assume represents a perfect fluid, and

$$\theta_{\mu\nu} = \frac{6}{\sigma} S_{\mu\nu} + \frac{1}{8\pi} \mathcal{E}_{\mu\nu} \,, \tag{5.24}$$

where the condition (1.20) has been used. In more details, we can write

$$\theta_0^{\,0} = \frac{1}{\sigma} \left(\frac{\rho^2}{2} + \frac{6}{k^4} \mathcal{U} \right) \tag{5.25}$$

$$\theta_1^{\,1} = -\frac{1}{\sigma} \left(\frac{\rho^2}{2} + \rho\, p + \frac{2}{k^4} \mathcal{U} \right) - \frac{4}{k^4} \frac{\mathcal{P}}{\sigma} \tag{5.26}$$

$$\theta_2^{\,2} = -\frac{1}{\sigma} \left(\frac{\rho^2}{2} + \rho\, p + \frac{2}{k^4} \mathcal{U} \right) + \frac{2}{k^4} \frac{\mathcal{P}}{\sigma} \,. \tag{5.27}$$

Therefore, the "quasi-Einstein" system (5.19)–(5.21) for the BW reads

$$k^2 \left(\frac{\rho^2}{2} + \frac{6\mathcal{U}}{k^4} \right) = -\frac{f^*}{r^2} - \frac{f^{*'}}{r} \tag{5.28}$$

$$k^2 \left[\left(\frac{\rho^2}{2} + \rho\, p + \frac{2}{k^4} \mathcal{U} \right) + \frac{4\mathcal{P}}{k^4} \right] = f^* \left(\frac{1}{r^2} + \frac{\xi'}{r} \right), \tag{5.29}$$

$$k^2 \left[\left(\frac{\rho^2}{2} + \rho\, p + \frac{2}{k^4} \mathcal{U} \right) - \frac{2\mathcal{P}}{k^4} \right] = \frac{1}{4} \left[f^* \left(2\,\xi'' + \xi'^2 + 2\frac{\xi'}{r} \right) \right.$$
$$\left. + f^{*'} \left(\xi' + \frac{2}{r} \right) \right], \tag{5.30}$$

where we have identified $\alpha = \sigma^{-1}$. On the other hand, the conservation equation (5.22) reads

$$k^4 \left(\rho\, \rho' + \rho'\, p + \rho\, p' \right) + 2 \left(\mathcal{U}' + 2\,\mathcal{P}' \right)$$
$$+ \frac{\nu'}{2} \left[k^4 \rho \left(\rho + p \right) + 4 \left(2\mathcal{U} + \mathcal{P} \right) \right] + \frac{12\,\mathcal{P}}{r} = 0, \tag{5.31}$$

which can be simplified by using the conservation equation for the perfect fluid in Eq. (1.30), yielding

$$\mathcal{U}' + 2\,\mathcal{P}' + \xi' \left(2\mathcal{U} + \mathcal{P} \right) + \frac{6\,\mathcal{P}}{r} = -\frac{k^4}{2} \left(\rho + p \right) \rho'. \tag{5.32}$$

Let us recall that the conservation equation (5.32) is a linear combination of the equations of motion for the BW gravitational sector, namely, the expressions in Eqs. (5.28)–(5.30). We see that after the decoupling, we obtain three independent equations, namely, the system (5.28)–(5.30), to determine three unknown functions $\{\mathcal{U}, \mathcal{P}, f^*\}$. This means that every perfect fluid configuration $\{p, \rho, \mu, \xi\}$ will have a specific BW solution determined by the MGD decoupling.

In this respect, a question naturally arises regarding the BW vacuum defined by $p = \rho = 0$ with $\mathcal{U}$ and $\mathcal{P}$ to be determined: what is the deformed exterior Schwarzschild solution obtained from the MGD decoupling? To answer this question, we impose the vacuum condition $p = \rho = 0$ on the system (5.28)–(5.30). This system then implies a first order differential equation for the deformation f^*, namely,

$$\left(\frac{\xi'}{2} + \frac{2}{r} \right) (f^*)' + \left(\xi'' + \frac{\xi'^2}{2} + \frac{2\,\xi'}{r} + \frac{2}{r^2} \right) f^* = 0 \tag{5.33}$$

which is solved by

$$f^* = C\, e^{-I(r)}, \tag{5.34}$$

where C is a constant and $I(r)$ the integral defined in Eq. (1.42) with $\nu = \xi$. Using the Schwarzschild solution (1.77) with $\mathcal{M} = M$ in Eq. (5.34) we obtain

$$f^* = \left(1 - \frac{2\,M}{r}\right) \frac{\ell_c}{2\,r - 3\,M} \,, \tag{5.35}$$

where the constant C in Eq. (5.34) was absorbed in the length scale $\ell_c \sim \sigma^{-1}$. Therefore the minimally deformed radial metric component reads

$$e^{-\lambda} = \left(1 - \frac{2\,M}{r}\right) \left(1 + \frac{\ell_c}{2\,r - 3\,M}\right) . \tag{5.36}$$

Using the geometric deformation (5.35) in the "quasi-Einstein" system (5.19)–(5.21), we obtain

$$\frac{6\,\mathcal{U}}{k^2} = -\frac{\ell_c\,M}{r^2(2\,r - 3\,M)^2} \tag{5.37}$$

and

$$\frac{6\,\mathcal{P}}{k^2} = -\frac{\ell_c\,(3\,r - 4\,M)}{r^2(2\,r - 3\,M)^2} . \tag{5.38}$$

The temporal metric component of the Schwarzschild solution (1.77) along with the deformed radial metric component (5.36) represents a hairy black-hole solution with outer horizon $r_H = 2\,M$, and primary hairs represented by the parameter ℓ. This solution was found in Ref. [47], and rediscovered in Ref. [17] in a more general scenario, where it was shown that this solution appears as a consequence of a conformal symmetry. In fact, it is associated with a traceless energy-momentum tensor, as in fact is the case of the tensor $\theta_{\mu\nu}$ in the vacuum [see Eqs. (5.25)–(5.27) for $p = \rho = 0$].

The solution (5.36) is the only one generated by the MGD. This result clearly shows the limitation of the MGD decoupling, since we know there are more black hole solutions in the BW context [48–50]. Therefore to find new black hole solutions in the BW we need to implement a generalisation of the MGD, or an extension of the MGD decoupling, as that developed in Ref. [2], which is beyond the goal of this book. Instead, we will see in the next section that it is still possible to generate new black hole solutions in the BW by the MGD decoupling when the vacuum is filled not only with the BW fields $\mathcal{U}$ and $\mathcal{P}$ but also with a generic source $\theta_{\mu\nu}$ we can think is related to a non-empty bulk, as implemented in Ref. [17].

5.2.1   Black Holes

Let us start by assuming the total energy-momentum tensor $\tilde{T}_{\mu\nu}$ in Eq. (5.1) only contains BW contributions, hence

$$\tilde{T}_{\mu\nu} = \frac{1}{8\pi}\,\mathcal{E}_{\mu\nu} + \theta_{\mu\nu} \equiv T_{\mu\nu} + \theta_{\mu\nu} \,, \tag{5.39}$$

where the energy-momentum tensor $T_{\mu\nu}$ is now given by the Weyl contribution on the brane. Therefore after decoupling the system (5.5)–(5.7) we end with (i) the Einstein equations for a pure BW sector $T_{\mu\nu} \sim \mathcal{E}_{\mu\nu}$ in Eqs. (5.15)–(5.17) to determine the metric functions ξ and μ, and (ii) the field equations with the source $\theta_{\mu\nu}$ in Eqs. (5.19)–(5.21) to determine the deformation f^*. In fact, we can use the latter equations to fix the (unknown) source term $\theta_{\mu\nu}$ as well.

Of all known black holes in the BW, here we will only deform the tidally charged solution (2.21). Its deformed version under the MGD decoupling reads

$$ds^2 = \left(1 - \frac{2M}{r} - \frac{q}{r^2}\right) dt^2 - \frac{dr^2}{1 - \frac{2M}{r} - \frac{q}{r^2} + \alpha f^*} - r^2 d\Omega^2 . \tag{5.40}$$

If we do not assume a priori a specific source $\theta_{\mu\nu}$, the system (5.19)–(5.21) contains four unknown functions to determine both $\theta_{\mu\nu}$ and f^*. Hence, we need to prescribe one additional condition to uniquely specify the solution. Next we will demand some physically motivated restriction on the energy-momentum tensor $\theta_{\mu\nu}$.

Let us start by considering the case of isotropic pressure, so that

$$\theta_1{}^1 = \theta_2{}^2 = \theta_3{}^3 . \tag{5.41}$$

Equations (5.20) and (5.21) then yield the differential equation for the MGD function

$$f^{*\prime} \left(\xi' + \frac{2}{r}\right) + f^* \left(2\,\xi'' + \xi'^2 - 2\frac{\xi'}{r} - \frac{4}{r^2}\right) = 0 . \tag{5.42}$$

After we replace the temporal metric component ξ of the metric shown in (5.40) into (5.42), we find the general solution is given by

$$f^* = \left(1 - \frac{2M}{r} - \frac{q}{r^2}\right) \left(\frac{r}{\ell_{\text{iso}}}\right)^2 e^{\frac{4q}{Mr}} \left(1 - \frac{M}{r}\right)^{2 + \frac{4q}{M^2}} , \tag{5.43}$$

where ℓ_{iso} is a constant with the dimensions of a length. Hence, the MGD radial component for an isotropic deformation of the tidally charged exterior becomes

$$e^{-\lambda} = e^\xi + \alpha f^*$$
$$= \left(1 - \frac{2M}{r} - \frac{q}{r^2}\right) \left[1 + \alpha \left(\frac{r}{\ell_{\text{iso}}}\right)^2 e^{\frac{4q}{Mr}} \left(1 - \frac{M}{r}\right)^{2 + \frac{4q}{M^2}}\right] , \tag{5.44}$$

which is clearly not asymptotically flat for $r \gg M$. We therefore conclude that the additional source $\theta_{\mu\nu}$ cannot contain an isotropic pressure if we wish to preserve asymptotic flatness.

Another possibility is that that the source $\theta_{\mu\nu}$ is associated with a conformal gravitational sector. Since the energy-momentum tensor for a conformally symmetric source must be traceless, we have

$$2\,\theta_2^2 = -\theta_0^{\ 0} - \theta_1^{\ 1}\,,\tag{5.45}$$

so that the system (5.19)–(5.21) becomes

$$-k^2\,\theta_0^{\ 0} = \frac{f^*}{r^2} + \frac{f^{*\prime}}{r}\tag{5.46}$$

$$-k^2\,\theta_1^{\ 1} = f^*\left(\frac{1}{r^2} + \frac{\xi'}{r}\right)\,,\tag{5.47}$$

where f^* is again the MGD function and ξ the undeformed tidally charged function shown in (5.40). From Eq. (5.45), we now find the radial deformation must satisfy the differential equation

$$f^{*\prime}\left(\frac{\xi'}{2} + \frac{2}{r}\right) + f^*\left(\xi'' + \frac{\xi'^2}{2} + 2\frac{\xi'}{r} + \frac{2}{r^2}\right) = 0\,,\tag{5.48}$$

and it is important to highlight that the conservation equation (5.22) remains a linear combination of the system (5.45)–(5.47). The general solution for Eq. (5.48) is given by

$$f^* = \left(1 - \frac{2\,M}{r} - \frac{q}{r^2}\right)\frac{\ell_c}{\sqrt{r\,(2\,r - 3\,M)} - q}\,e^{\frac{3\,M\,\arctan\left(\frac{3M-4r}{\sqrt{-9\,M^2-8q}}\right)}{\sqrt{-9\,M^2-8q}}}\,,\tag{5.49}$$

with ℓ_c again a constant with units of a length. Thus the conformally deformed tidally charged radial metric component becomes

$$e^{-\lambda} = \left(1 - \frac{2\,M}{r} - \frac{q}{r^2}\right)\left[1 + \frac{\ell_c}{\sqrt{r\,(2\,r - 3\,M)} - q}\,e^{\frac{3\,M\,\arctan\left(\frac{3M-4r}{\sqrt{-9\,M^2-8q}}\right)}{\sqrt{-9\,M^2-8q}}}\right]\,,\tag{5.50}$$

where $\ell = \alpha\,\ell_c$. This solution represents a black hole with the same two horizons of the original tidally charged solution in Eq. (2.21), namely, $r_{\rm H} = M \pm \sqrt{M^2 + q}$.

Finally, let us consider a source $\theta_{\mu\nu}$ which satisfies the condition of null tangential pressure, namely

$$\theta_2^2 = 0\,,\tag{5.51}$$

which, according to Eq. (5.21), yields

$$f^{*\prime}\left(\xi' + \frac{2}{r}\right) + f^*\left(2\,\xi'' + \xi'^2 + 2\frac{\xi'}{r}\right) = 0\,,\tag{5.52}$$

whose solution is given by

$$f^* = C \left(1 - \frac{2M}{r} - \frac{q}{r^2} \right) \left(1 - \frac{M}{r} \right)^{\frac{2q}{M^2}} e^{\frac{2q}{Mr}} , \qquad (5.53)$$

with C a constant. Hence the deformed tidally charged solution becomes

$$e^{-\lambda} = \left(1 - \frac{2M}{r} - \frac{q}{r^2} \right) \left[1 + C \left(1 - \frac{M}{r} \right)^{\frac{2q}{M^2}} e^{\frac{2q}{Mr}} \right] , \qquad (5.54)$$

which is not asymptotically flat for $r \gg M$. We therefore conclude that the source $\theta_{\mu\nu}$ must contain a non-null tangential pressure if we wish to preserve asymptotic flatness.

5.2.2   Interior Solutions

We can also find interior solutions for a self-gravitating system in the same way as we determined exterior solutions. The deformed interior metric under the MGD decoupling reads

$$ds^2 = \left(1 - \frac{2m(r)}{r} \right) dt^2 - \frac{dr^2}{1 - \frac{2m(r)}{r} + \alpha f^*} - r^2 \, d\Omega^2 , \qquad (5.55)$$

and we recall that after the decoupling, we end up with the three independent Eqs. (5.28)–(5.30) which we can use to determine the three unknown functions $\mathcal{U}$, $\mathcal{P}$ and f^*. By combining Eqs. (5.28)–(5.30) we find the first order differential equation for the function f^*, given by

$$\left(\frac{\xi'}{2} + \frac{2}{r} \right) (f^*)' + \left(\xi'' + \frac{\xi'^2}{2} + \frac{2\,\xi'}{r} + \frac{2}{r^2} \right) f^* = k^2 \, (\rho^2 + 3\rho\, p) \qquad (5.56)$$

whose formal solution is

$$f^* = k^2 \, e^{-I(r)} \int_0^r \frac{e^{I(x)} \, \rho \, (\rho + 3\, p)}{\xi'/2 + 2/x} \, dx + C \, e^{-I(r)} , \qquad (5.57)$$

where C a constant and $I(r)$ the integral defined in Eq. (1.42) with $\nu = \xi$. We see that, in the vacuum $\rho = p = 0$, the expression (5.57) yields the one in (5.34). However, for interior solutions, the integration constant C must be zero to avoid the singularity in the origin $r = 0$. Also notice that the solution (5.57) is the same as that in (1.60) [with $C = 0$ and $\sigma^{-1} = \alpha$ considered in the transformation (5.14)]. Now we can choose *any* spherically symmetric and static solution to the Einstein equations (5.15)–(5.17) to generate its BW version. For instance, we can pick up the Tolman IV perfect fluid solution shown in Eqs. (2.74)–(2.77) [where $\nu_{(-)} = \xi$, $e^{-\lambda_{(-)}} = \mu$ and $f^* = 0$].

This perfect fluid configuration is a solution of the system (5.15)–(5.18), and will generate a specific BW solution via the MGD decoupling. Pluging the expressions in Eqs. (2.74), (2.76) and (2.77) into Eq. (5.57) we will reproduce the expression in Eq. (2.79). It should be by now clear that every spherically symmetric solution will have a unique BW version.

5.2.3 Beyond Perfect Fluid Solutions by MGD Decoupling in the Brane-World

Whereas the MGD approach is a powerful method to investigate BW effects inside self-gravitating configurations, it also suffers from the limitation that it can only be used to generate the BW version of perfect fluid GR solutions. However, as we have already mentioned, the gravitational decoupling can be applied to any kind of gravitational source, no matter its nature. We will then show how to generate the BW version of any anisotropic solution in GR, and discuss a specific example.

Let us begin by considering an anisotropic fluid with

$$T_{\mu\nu} = (\rho + p_t)\, u_\mu\, u_\nu - p_t\, g_{\mu\nu} + (p_r - p_t)\chi_\mu\chi_\nu\, , \tag{5.58}$$

where p_r and p_t are the radial and tangential pressures respectively, $u^\mu = e^{-\nu/2}\,\delta_0^\mu$ is the fluid 4-velocity and $\chi^\mu = e^{-\lambda/2}\,\delta_1^\mu$ a unit radial vector field, both defined in the Schwarzschild-like coordinates of the metric (1.4). These two vectors must satisfy

$$u^\mu u_\mu = -\chi^\mu\chi_\mu = 1\, , \quad u^\mu\chi_\mu = 0\, . \tag{5.59}$$

Using the expression (5.58) in the high-energy corrections (1.17), we find that the Einstein field equations (1.27)–(1.29) become

$$k^2\left[\rho + \frac{1}{\sigma}\left(\frac{\rho^2}{2} - \frac{(p_r - p_t)^2}{2}\right) + \frac{6}{k^4}\frac{\mathcal{U}}{\sigma}\right] = \frac{1}{r^2} - e^{-\lambda}\left(\frac{1}{r^2} - \frac{\lambda'}{r}\right) \tag{5.60}$$

$$k^2\left[p_r + \frac{1}{\sigma}\left(\frac{\rho^2}{2} + \rho\,p_t + \frac{p_t^2 - p_r^2}{2}\right) + \frac{2}{k^4}\frac{\mathcal{U}}{\sigma} + \frac{4}{k^4}\frac{\mathcal{P}}{\sigma}\right]$$
$$= -\frac{1}{r^2} + e^{-\lambda}\left(\frac{1}{r^2} + \frac{\nu'}{r}\right) \tag{5.61}$$

$$k^2\left[p_t + \frac{1}{\sigma}\left(\frac{\rho}{2}(\rho + p_r + p_t) + \frac{p_r}{2}(p_r - p_t)\right) + \frac{2}{k^4}\frac{\mathcal{U}}{\sigma} - \frac{2}{k^4}\frac{\mathcal{P}}{\sigma}\right]$$
$$= \frac{1}{4}e^{-\lambda}\left(2\,\nu'' + \nu'^2 - \lambda'\,\nu' + 2\,\frac{\nu' - \lambda'}{r}\right)\, . \tag{5.62}$$

As a check, we can see that the system (5.60)–(5.62) reduces to (1.27)–(1.29) for the isotropic case $p_r = p_t = p$. Now, following the MGD decoupling approach, we compare the system (5.5)–(5.7) with (5.60)–(5.62) to identify $T_{\mu\nu}$ with the anisotropic system (5.15)–(5.18) and $\theta_{\mu\nu}$ with the BW terms, given by

$$\theta_0^{\,0} = \frac{1}{\sigma}\left(\frac{\rho^2}{2} - \frac{(p_r - p_t)^2}{2}\right) + \frac{6\mathcal{U}}{\sigma k^4} \tag{5.63}$$

$$\theta_1^{\,1} = -\frac{1}{\sigma}\left(\frac{\rho^2}{2} + \rho\, p_t + \frac{p_t^2 - p_r^2}{2}\right) - \frac{2\mathcal{U}}{\sigma k^4} - \frac{4\,\mathcal{P}}{k^4\,\sigma} \tag{5.64}$$

$$\theta_2^{\,2} = -\frac{1}{\sigma}\left(\frac{\rho}{2}(\rho + p_r + p_t) + \frac{p_r}{2}(p_r - p_t)\right) - \frac{2\mathcal{U}}{\sigma k^4} + \frac{2\,\mathcal{P}}{k^4\,\sigma}\,. \tag{5.65}$$

Therefore, the "quasi-Einstein" system (5.19)–(5.21) becomes

$$k^2\left[\frac{\rho^2}{2} - \frac{(p_r - p_t)^2}{2} + \frac{6\mathcal{U}}{k^4}\right] = -\frac{f^*}{r^2} - \frac{f^{*\prime}}{r} \tag{5.66}$$

$$k^2\left[\frac{\rho^2}{2} + \rho\, p_t + \frac{p_t^2 - p_r^2}{2} + \frac{2\mathcal{U}}{k^4} + \frac{4\mathcal{P}}{k^4}\right] = f^*\left(\frac{1}{r^2} + \frac{\xi'}{r}\right) \tag{5.67}$$

$$k^2\left[\frac{\rho}{2}(\rho + p_r + p_t) + \frac{p_r}{2}(p_r - p_t) + \frac{2\mathcal{U}}{k^4} - \frac{2\mathcal{P}}{k^4}\right] = \frac{1}{4}\left[f^*\left(2\xi'' + \xi'^2 + 2\frac{\xi'}{r}\right)\right.$$
$$\left. + f^{*\prime}\left(\xi' + \frac{2}{r}\right)\right]\,, \tag{5.68}$$

where we have identified $\alpha = \sigma^{-1}$. On the other hand, the conservation equation (5.22) reads

$$(\rho' + p_t')(\rho + p_t) - p_r p_r' + \frac{2\mathcal{U}'}{k^4} + \frac{4\mathcal{P}'}{k^4}$$
$$+ \frac{\xi'}{2}\left(\rho^2 - p_r^2 + p_r p_t + \rho\, p_t + \frac{8\mathcal{U}}{k^4} + \frac{4\mathcal{P}}{k^4}\right)$$
$$+ \frac{2}{r}\left(\frac{\rho\, p_t}{2} + \frac{p_t^2}{2} - p_r^2 - \frac{\rho\, p_r}{2} + \frac{p_r p_t}{2} + \frac{6\mathcal{P}}{k^4}\right) = 0\,, \tag{5.69}$$

which can be simplified by using the conservation equation for the anisotropic fluid $T_\mu^{\,\nu} = \text{diag}(\rho, -p_r, -p_t, -p_t)$ in Eq. (5.18), yielding

$$\mathcal{U}' + 2\mathcal{P}' + \xi'(2\mathcal{U} + \mathcal{P}) + \frac{6\mathcal{P}}{r} = -\frac{k^4}{2}(\rho' + p_t')(\rho + p_t)$$
$$- \frac{k^4}{2r}(\rho + p_t)(p_t - p_r) - k^4\frac{\xi'}{4}\left[\rho^2 + p_r p_t + \rho(p_r + p_t)\right]\,. \tag{5.70}$$

We can see that the conservation equation (5.70) reproduces Eq. (5.32) for the isotropic case $p_r = p_t = p$ [after using the conservation equation for a perfect fluid in (1.30), with $\nu = \xi$]. Notice that now, beside the gradients of the density, the anisotropy is also a source for the fields $\mathcal{U}$ and $\mathcal{P}$. Let us recall that the conservation equation (5.70) is a linear combination of the equations of motion for the BW gravitational sector described by Eqs. (5.66)–(5.68). We see that after the decoupling, we end with the three independent Eqs. (5.66)–(5.68) which we can use to determine the three unknown functions $\mathcal{U}$, $\mathcal{P}$ and f^*. This means that every given anisotropic configuration $\{\rho, p_r, p_t, \mu, \xi\}$ will generate a specific BW solution via the MGD decoupling.

By combining Eqs. (5.66)–(5.68) we find the first order differential equation

$$\left(\frac{\xi'}{2} + \frac{2}{r}\right)(f^*)' + \left(\xi'' + \frac{\xi'^2}{2} + \frac{2\,\xi'}{r} + \frac{2}{r^2}\right) f^*$$
$$= k^2 \left[\rho^2 + \rho(p_r + 2p_t) + (p_r - p_t)^2\right], \tag{5.71}$$

whose formal solution is given by

$$f^* = k^2 e^{-I(r)} \int_0^r \frac{e^{I(x)} \left[\rho^2 + \rho(p_r + 2p_t) + (p_r - p_t)^2\right]}{\xi'/2 + 2/x} dx + C e^{-I(r)}, \tag{5.72}$$

where C is a constant and $I(r)$ the integral defined in Eq. (1.42) with $\nu = \xi$. Notice that the geometric deformation f^* in Eq. (5.72) reduces to the one in Eq. (5.57) for the isotropic case $p_r = p_t = p$. Next we will find the BW version of a specific interior anisotropic solution.

Let us begin by considering a compact self-gravitating system sustained only by tangential stresses described by

$$e^\xi = B^2 \left(1 + \frac{r^2}{A^2}\right), \tag{5.73}$$

$$e^{-\mu} = \frac{A^2 + r^2}{A^2 + 3\,r^2}, \tag{5.74}$$

$$\rho = \frac{6\left(A^2 + r^2\right)}{k^2\left(A^2 + 3\,r^2\right)^2}, \tag{5.75}$$

$$p_t = \frac{3\,r^2}{k^2\left(A^2 + 3\,r^2\right)^2}, \tag{5.76}$$

$$p_r = 0, \tag{5.77}$$

where $0 \le r \le R_s$ and $r = R_s$ defines the surface of the compact object. A direct interpretation of this class of solutions (albeit not unique, as pointed out in Ref. [45]) is in terms of a cluster of particles moving in randomly oriented circular orbits [51]. The constants A and B can be determined from the matching conditions between this interior solution and the exterior metric for $r > R_s$.

The configuration $\{p_r, p_t, \rho, \mu, \xi\}$ in Eqs. (5.73)–(5.77) is a solution of the system (5.15)–(5.18), and will generate a specific BW solution via the MGD decoupling. On using Eqs. (5.73) and (5.75)–(5.77) in Eq. (5.72) we find

$$f^*(r) = \frac{C\,(A^2 + r^2)}{r\,(2A^2 + 3r^2)^{3/2}} + \frac{(A^2 + r^2)}{k^2\,r\,(2A^2 + 3r^2)^{3/2}}$$

$$\times \left[\sqrt{3}\log\left(\sqrt{6A^2 + 9r^2} + 3r\right) + \sqrt{3}\arctan\left(\frac{r}{\sqrt{\frac{2A^2}{3} + r^2}}\right) \right.$$

$$\left. -\frac{3r\sqrt{2\,A^2 + 3r^2}\,(2A^4 + 7A^2r^2 + 6\,r^4)}{2\,(A^2 + 3r^2)^3} \right] , \qquad (5.78)$$

where C is a constant. Now, in order to avoid a singularity in the origin $r = 0$, this
constant must take the value

$$C = -\frac{\sqrt{3}}{k^2}\log(\sqrt{6\,A^2}) . \qquad (5.79)$$

From the "quasi-Einstein" equations (5.66) and (5.67) we obtain the Weyl fields $\mathcal{U}$
and $\mathcal{P}$ in terms of the geometric deformation as

$$\frac{6\mathcal{U}}{k^2} = -\frac{f^*}{r^2} - \frac{(f^*)'}{r} - \frac{k^2}{2}\left[\rho^2 + (p_r - p_t)^2\right] , \qquad (5.80)$$

$$\frac{6\mathcal{P}}{k^2} = -\frac{(f^*)'}{2}\left(\frac{\xi'}{2} + \frac{1}{r}\right) - \frac{f^*}{2}\left(\xi'' + \frac{\xi'^2}{2} - \frac{\xi'}{r} - \frac{2}{r^2}\right)$$

$$+ \frac{k^2}{2}(p_r - p_t)(\rho + 2\,p_r - p_t) . \qquad (5.81)$$

Figures 5.1 and 5.2 show the behaviour of the Wayl fields $\mathcal{U}$ and $\mathcal{P}$ for different
compactness. We see that extra-dimensional effects are larger for more compact
distributions.

Fig. 5.1 Scalar Weyl
function $\mathcal{U}(r)$ for three
different compactness M/R_s

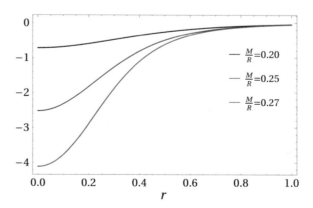

Fig. 5.2 Weyl function
$[\mathcal{P} \times 10]$ for three different
compactness M/R_{s}

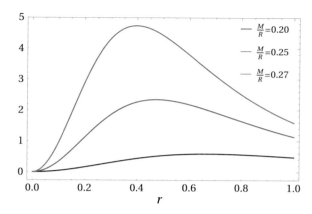

Fig. 5.3 Effective radial
pressure $\tilde{p}_r$ for the GR case
$(p_r = 0)$ and the BW

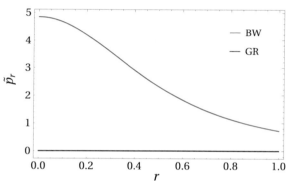

> **Extra-dimensional effects are always stronger the larger the compactness of the source.**

On the other hand, Fig. 5.3 shows the effective radial pressure $\tilde{p}_r$, which according to (5.61) reads

$$\tilde{p}_r = p_r + \frac{1}{\sigma}\left(\frac{\rho^2}{2} + \rho\, p_t + \frac{p_t^2 - p_r^2}{2}\right) + \frac{2}{k^4}\frac{\mathcal{U}}{\sigma} + \frac{4}{k^4}\frac{\mathcal{P}}{\sigma}. \qquad (5.82)$$

In this particular case, with $p_r = 0$ for the GR solution (5.73)–(5.77), we see that the effective radial pressure (5.82) is entirely due to extra-dimensional effects.

Epilogue

We would like to end this book by emphasising the importance of what we have shown in this final Chapter: The Gravitational Decoupling approach represents a powerful tool, which allows us to deal with the Einstein field equations in a novel and above all very pragmatic way. Its applications, so far, have shown its effectiveness for the study of black holes, the interior of self-gravitating objects, charged stellar distributions, the Einstein-Klein Gordon system, the case of gravity in $(2 + 1)$-dimensions, extra-dimensional theories, other theories beyond GR, and more [3–43]. We therefore believe that the extension of this method to more complex scenarios, such as non-static or less symmetrical cases, represents a stimulating challenge for future research.

References

1. J. Ovalle, Decoupling gravitational sources in GR: from perfect to anisotropic fluids. Phys. Rev. D **95**, 104019 (2017)
2. J. Ovalle, Decoupling gravitational sources in GR: the extended case. Phys. Lett. B **787** (2019)
3. C. Las Heras, P. Leon, Using MGD gravitational decoupling to extend the isotropic solutions of Einstein equations to the anisotropical domain. Fortschr. Phys. 1800036 (2018)
4. A. Fernandes-Silva, R. da Rocha, Gregory-Laflamme analysis of MGD black strings. Eur. Phys. J. C **78**, 271 (2018)
5. J. Ovalle, R. Casadio, R. da Rocha, A. Sotomayor, Anisotropic solutions by gravitational decoupling. Eur. Phys. J. C **78**, 122 (2018)
6. L. Gabbanelli, A. Rincón, C. Rubio, Gravitational decoupled anisotropies in compact stars. Eur. Phys. J. C **78**, 370 (2018)
7. E. Contreras, P. Bargueño, Minimal geometric deformation decoupling in 2 + 1 dimensional space-times. Eur. Phys. J. C **78**, 558 (2018)
8. M. Sharif, S. Sadiq, Gravitational decoupled anisotropic solutions for cylindrical geometry. Eur. Phys. J. Plus **133**, 245 (2018)
9. M. Sharif, S. Sadiq, Gravitational decoupled charged anisotropic spherical solutions. Eur. Phys. J. C **78**, 410 (2018)
10. Grigoris Panotopoulos, Ángel Rincón, Minimal geometric deformation in a cloud of strings. Eur. Phys. J. C **78**, 851 (2018)
11. M. Sharif, S. Saba, Gravitational decoupled anisotropic solutions in $f(\mathcal{G})$ gravity. Eur. Phys. J. C **78**, 921 (2018)
12. Ernesto Contreras, Pedro Bargue, Minimal geometric deformation in asymptotically (A-)dS space-times and the isotropic sector for a polytropic black hole. Eur. Phys. J. C **79**, 985 (2018)
13. Milko Estrada, Francisco Tello-Ortiz, A new family of analytical anisotropic solutions by gravitational decoupling. Eur. Fis. J. Plus **133**, 453 (2018)
14. E. Morales, F. Tello-Ortiz, Charged anisotropic compact objects by gravitational decoupling. Eur. Phys. J. C **78**(8), 618 (2018)
15. R. Casadio, P. Nicolini, R. da Rocha, GUP Hawking fermions from MGD black holes. Class. Quant. Grav. **35**, 185001 (2018)
16. A. Fernandes-Silva, A.J. Ferreira-Martins, R. da Rocha, The extended minimal geometric deformation of SU(N) dark glueball condensates. Eur. Phys. J. C **78**, 631 (2018)
17. J. Ovalle, R. Casadio, R. da Rocha, A. Sotomayor, Z. Stuchlik, Black holes by gravitational decoupling. Eur. Phys. J. C **78**, 960 (2018)

18. E. Contreras, Minimal geometric deformation: the inverse problem. Eur. Phys. J. C **78**, 678 (2018)
19. R. Pérez, A new anisotropic solution by MGD gravitational decoupling. Eur. Phys. J. Plus **133**, 244 (2018). [Erratum: Eur. Phys. J. Plus **134**(7), 369 (2019)]
20. E. Morales, F. Tello-Ortiz, Compact anisotropic models in general relativity by gravitational decoupling. Eur. Phys. J. C **78**, 841 (2018)
21. M. Estrada, R. Prado, The gravitational decoupling method: the higher dimensional case to find new analytic solutions. Eur. Phys. J. Plus **134**,168 (2019)
22. J. Ovalle, A. Sotomayor, A simple method to generate exact physically acceptable anisotropic solutions in general relativity. Eur. Phys. J. Plus **133**, 428 (2018)
23. G. Panotopoulos, Á. Rincón, Minimal geometric deformation in a cloud of strings. Eur. Phys. J. C **78**, 851 (2018)
24. J. Ovalle, R. Casadio, R. da Rocha, A. Sotomayor, Z. Stuchlik, Einstein-Klein-Gordon system by gravitational decoupling. EPL **124**, 20004 (2018)
25. Ernesto Contreras, Gravitational decoupling in 2 + 1 dimensional space-times with cosmological term. Class. Quant. Grav. **36**, 095004 (2019)
26. A. Fernandes-Silva, A.J. Ferreira-Martins, R. da Rocha, Extended quantum portrait of MGD black holes and information entropy. Phys. Lett. B **791**, 323 (2019)
27. S.K. Maurya, F. Tello-Ortiz, Generalized relativistic anisotropic compact star models by gravitational decoupling. Eur. Phys. J. C **79**, 85 (2019)
28. E. Contreras, A. Rincón, P. Bargueño, A general interior anisotropic solution for a BTZ vacuum in the context of the minimal geometric deformation decoupling approach. Eur. Phys. J. C **79**, 216 (2019)
29. E. Contreras, P. Bargueño, Extended gravitational decoupling in 2 + 1 dimensional space-times. Class. Quant. Grav. **36**, 215009 (2019)
30. L. Gabbanelli, J. Ovalle, A. Sotomayor, Z. Stuchlik, R. Casadio, A causal Schwarzschild-de Sitter interior solution by gravitational decoupling. Eur. Phys. J. C **79**, 486 (2019)
31. M. Estrada, A way of decoupling gravitational sources in pure Lovelock gravity. Eur. Phys. J. C **79**, 918 (2019)
32. J. Ovalle, C. Posada, Z. Stuchlik, Anisotropic ultracompact Schwarzschild star by gravitational decoupling. Class. Quant. Grav. **36**, 205010 (2019)
33. S. K. Maurya F. Tello-Ortiz, Charged anisotropic compact star in $f(R, \mathcal{T})$ gravity: a minimal geometric deformation gravitational decoupling approach. arXiv:1905.13519 [gr-qc]
34. M. Sharif, A. Waseem, Effects of charge on gravitational decoupled anisotropic solutions in $f(R)$ gravity. Chin. J. Phys. **60**, 426 (2019)
35. S. Hensh, Z. Stuchlik, Anisotropic Tolman VII solution by gravitational decoupling. Eur. Phys. J. C **79**, 834 (2019)
36. F.X. Cedeño, E. Contreras, Gravitational decoupling in cosmology. arXiv:1907.04892 [gr-qc]
37. P. León, A. Sotomayor, Braneworld Gravity under gravitational decoupling. Fortschr. Phys. **1900077** (2019)
38. V.A. Torres-Snchez, E. Contreras, Anisotropic neutron stars by gravitational decoupling. Eur. Phys. J. C **79**, 829 (2019)
39. A. Rincón, L. Gabbanelli, E. Contreras, F. Tello-Ortiz, Minimal geometric deformation in a Reissner-Nordstrom background. Eur. Phys. J. C **79**, 873 (2019)
40. M. Sharif, S. Sadiq, 2+1-dimensional gravitational decoupled anisotropic solutions. Chin. J. Phys. **60**, 279 (2019)
41. K.N. Singh, S.K. Maurya, M.K. Jasim, F. Rahaman, Minimally deformed anisotropic model of class one space-time by gravitational decoupling. Eur. Phys. J. C **79** 851 (2019)
42. M. Sharif, A. Waseem, Anisotropic spherical solutions by gravitational decoupling in $f(R)$ gravity. Ann. Phys. **405**, 14 (2019)
43. S.K. Maurya, A completely deformed anisotropic class one solution for charged compact star: a gravitational decoupling approach. Eur. Phys. J. C **79**, 958 (2019)
44. P. Burikham, T. Harko, M.J. Lake, Mass bounds for compact spherically symmetric objects in generalized gravity theories. Phys. Rev. D **94**(6), 064070 (2016)

45. L. Herrera, N.O. Santos, Local anisotropy in self-gravitating systems. Phys. Rep. **286**, 53 (1997)
46. M.K. Mak, T. Harko, Anisotropic stars in GR. Proc. R. Soc. Lond. A **459**, 393 (2003)
47. C. Germani, R. Maartens, Stars in the braneworld. Phys. Rev. D **64**, 124010 (2001)
48. N. Dadhich, R. Maartens, P. Papadopoulos, V. Rezania, Black holes on the brane. Phys. Lett. B **487**, 1–6 (2000)
49. R. Casadio, A. Fabbri, L. Mazzacurati, New black holes in the brane world? Phys. Rev. D **65**, 084040 (2002)
50. P. Figueras, T. Wiseman, Gravity and large black holes in Randall-Sundrum II braneworlds. Phys. Rev. Lett. **107**, 081101 (2011)
51. A. Einstein, Ann. Math. **40**, 922–936 (1939)

Printed in the United States
By Bookmasters